CLIMATE CHANGE AND LIFE

CLIMATE CHANGE AND LIFE

THE COMPLEX CO-EVOLUTION OF CLIMATE AND LIFE ON EARTH, AND BEYOND

Written by

GABRIEL FILIPPELLI

Elsevier
Radarweg 29, PO Box 211, 1000 AE Amsterdam, Netherlands
The Boulevard, Langford Lane, Kidlington, Oxford OX5 1GB, United Kingdom
50 Hampshire Street, 5th Floor, Cambridge, MA 02139, United States

ISBN: 978-0-12-822568-4

For information on all Elsevier publications visit our website at https://www.elsevier.com/books-and-journals

Publisher: Candice Janco
Acquisitions Editor: Jessica Mack
Editorial Project Manager: Joshua Mearns
Production Project Manager: Sruthi Satheesh
Cover Designer: Miles Hitchen

Typeset by TNQ Technologies

Contents

CHAPTER ONE

Earth over the past 4.5 billion years—a brief history

Introduction

Two numbers stand out in defining the Earth, and may in fact be critical to the evolution and existence of life everywhere it may reside in the Universe—0 and 100. You are familiar with them as the freezing point of liquid water, 0°C, and the boiling point of water, 100°C. It is no coincidence why these numbers are so, well, logical compared to the Fahrenheit system markers of 32 degrees and 212 degrees for the freezing and boiling points of water. The Celsius scale is actually defined by these very properties of water. But why are these markers considered so critical to life? It might be myopic, but it is because liquid water is essential for all biological functions on our planet, for all organisms. Every cell, every plant, and every elephant requires water to be in the liquid state in or near its surface to survive. We will explore the full range of life on our planet, which actually pushes those extremes in temperature but in rare and unique ways, in later chapters, but the presence of liquid water is so critical for basic biological functions that NASA has for decades shaped it's astrobiology program and exploration plans around the precept of "follow the water."

The Earth has been between those freezing and boiling barriers for at least four billion years, if not longer. It has swayed perilously close to both boundaries at times, but the global average temperature has stayed within the range of liquid water being present on the surface for that whole time. That means rivers, lakes, oceans, clouds, and rain have been a continuous feature of our planet. This is a key, and important, takeaway from this brief 0—100 lecture—namely, that the conditions were ripe for biological organisms to have survived on the surface of the planet for at least four billion years. And we have fossil evidence for life back nearly that far, thus confirming our water—life bias.

The Earth itself has existed in a solid, planetary form for 4.55 billion years, roughly a third of the lifetime of the Universe (Fig. 1.1).

Climate Change and Life
ISBN: 978-0-12-822568-4
https://doi.org/10.1016/B978-0-12-822568-4.00006-7

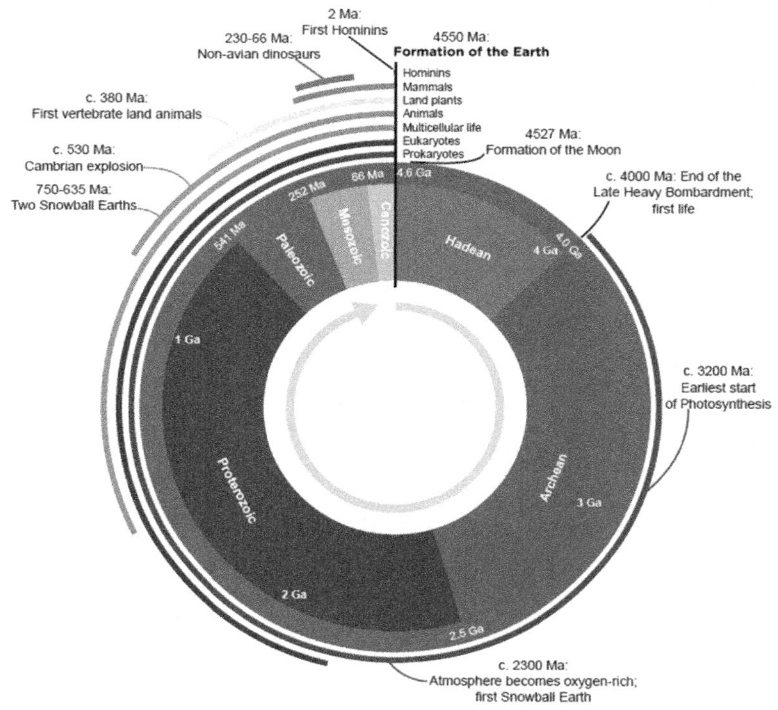

Figure 1.1 ***The History of Earth.*** The 4.5 billion (Ga) year history of Earth includes major geologic events as well as evolutionary ones, with more complex life emerging on the planet beginning roughly 600 million years ago.

You as an individual, or even our species, *Homo sapiens*, represent an unfathomably small fraction of that time. It seems difficult to fully understand, as our own perceptions of time are strongly shaped by our own individual existence, but here is the reality–a human life lasts less than 0.000002% of the lifetime of the Earth. Our species has lasted only 0.004%. That is an incredibly brief flicker of time with respect to the Earth. But nevertheless, we have an understanding of past events in our own lives, and those of our ancestors, and theirs. We record our own human history, through diaries, or books, or memories. The Earth has recorded its history through fossils and rocks. Our human histories become blurry with time–pages get torn, memories are lost. The Earth's history similarly gets blurry, as the rock record gets more and more discontinuous the further back in time you go. And not only that, it is also highly biased toward "remembering" major events. Think about the books that you read, or even your own memories. Does history record what a farmer ate in Mesopotamia thousands of years ago, on a Sunday? No, history records earthquakes, invasions, and coronations. Do you remember what you ate for lunch on February 18, 2009? No, but you remember your high school graduation and your Mom's birthday (well, you should, at least!). The point here is that Earth's history is a long one, we are mostly making guesses about events that happened in deep time (billions of years ago), and mostly only seeing extreme events in the geologic record, like mass extinctions, major changes in global chemistry, major asteroid impacts, etc. But there is some power in a record biased toward the extremes, because these intervals of time are critical markers between one state you as a high school student, and another, you as a graduate heading off to college.

This book focuses on the extremes, and indeed, takes advantage of the geologic bias to reveal how major disruptions in planetary systems have driven global climate changes and the evolution, destruction, and proliferation of life on this planet. The extremes will reveal perhaps one of the most profoundly humbling aspects of our little blue and white ball circling around an average star on the outskirts of our own Milky Way galaxy–the earth's incredible resilience through disasters, major and minor. These disasters include the major ones, such as the wholesale shift in the earth's atmospheric chemistry, globe-encompassing glacial events, and catastrophic asteroid collisions. They also include the "minor" ones that are nevertheless as important in the long term, such as the unique conditions that finally resulted in the evolution of multicellular organisms after likely millions of iterations and the coincidental cooccurrence of ice sheet retreat and maturation of the protective ozone layer around the Earth that resulted in the greatest evolutionary explosion of animals that has ever occurred on the planet.

In this chapter, we cover in depth the earliest formation of the Earth and the Moon, the various cycles of water, carbon, and geologic processes that dominated early Earth history, and the basic processes that drive the cycling of minerals and elements within our Earth system. These basic system processes are why the Earth has remained resilient in the face of major disasters and has harbored a vast and teaming set of organisms for most of its history, and why our solar system sisters, Venus and Mars, failed the test of resilience at various times in their history. The final part of this chapter provides a brief survey of the second half of the planet's existence. This is brief because the critical events that shaped the coevolution of climate and life are covered in depth in the rest of the book, even to current and future timelines, where the control of climate change is largely in our hands and the life around us is experiencing changes at rates heretofore unseen in earth history.

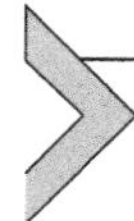

A brief history of planet earth

The first 500 million years

Our atomic clock

We have precisely dated the age of the Earth using the "clocks" that are embedded in elements or more precisely, the radioactive decay of versions of those elements. Elements themselves are defined strictly by the number of positive particles that reside in the nucleus, or middle, of an atom. Iron, for example, has 26 positive particles. If one positive particle (called a proton) is somehow added, iron becomes cobalt. If one is removed from iron, it becomes manganese. This number of protons in the nucleus, called the atomic number, defines the nature and state of that element and is the fundamental characteristic behind the most popular way to organize the elements on a chart, the so-called Periodic Table of the Elements, because it classifies groups of elements that behave similarly to each other in chemical, biological, or physical reactions.

Although the number of protons for a given element is immutable, the number of neutrons and the number of negatively charged particles surrounding the nucleus (electrons) can and do vary. For example, iron can have 26 electrons, or 28, or 29. It is still iron, but it reacts differently with other elements in these states. But iron also can occur is several isotopic states. For example, iron can have 28 neutrons or 30, 31, or 32. It largely behaves similarly no matter how many neutrons because all of the isotopes of iron are stable. Some elements, however, have transitory, atomically unstable isotopes. Uranium is a perfect example. Indeed, we use that instability for everything from producing energy in nuclear power plants to releasing energy in massive atomic bomb explosions to, far less dramatically to the average person but infinitely more exciting to geologists, dating the age of the Earth!

Uranium, with an elemental symbol of U, has three 92 protons but a dizzying number of isotopes, all unstable. It just so happens that the most abundant of these, U238, comprises over 99% of all U atoms on the planet, and like all unstable, radioactive isotopes decays over time in a predictable manner, becoming another element altogether because the decay involves loss of protons from the nucleus. In the case of U238, one measure of this decay rate, the half-life, is about 4.5 billion years. This means that half of the U238 that was originally present will have decayed away to something else in 4.5 billion years. Of course, you need to measure "something else" to which it has decayed away to know how much was there to start with, and in this case, we have excellent data on the eventual "daughter" products of that decay, often trapped in extremely resistant minerals like zircon. Interestingly, the decay of U238 not only allows us to date the age of the Earth, but the energy released from this decay is also responsible for much of the internal heat of the planet, which drives its active volcanic dynamics.

The Hadean Era

Sometimes, a name says it all. A case in point is the Hadean Era. This was the tumultuous first half billion years of Earth's history, from 4.55 to 4.0 billion years ago. Named for Hades, the Greek god of the underworld itself, it was an extremely hot, tumultuous, and largely unrecorded early period of chaos before some stability entered the solar system, and the Earth itself, around four billion years ago (A new picture of the early Earth, 2008). The beginning of the Hadean began eventfully, with what might have been a head-on collision between another large planetoid, Theia (another Greek reference), which was roughly the size of Mars and likely followed the early Earth in its orbital path. Theia may have been orbiting slightly faster than Earth, and then caught up and hit the Earth in a low-speed collision that effectively merged this planetoid with the Earth and ejected a large amount of material on the opposite of the impact locations. This ejected material coalesced from the forces of gravity and became our Moon. Many lines of evidence support this hypothesis, including the larger-than-expected size of the Earth's core and the similar mineral composition of the Moon and the Earth. But the long lens of history can indeed be unfocused, and new evidence or more advanced modeling might unearth alternative explanations.

Theia was not the only Earth collider. Numerous bodies big and small hit the Earth during this phase called the "early bombardment" interval, as the Earth swept a path through its orbital travel, slowly but surely exerting its gravitational pull as it orbited, like a magnet traveling through a scattering of iron filings. These collisions contributed additional water to the vast amount that was already present upon early Earth formation. They also contributed repeated episodes of collisional energy, causing the Earth to heat up each time a larger impactor struck. Surprisingly, there is evidence that liquid water, and indeed oceans, existed on Earth all the way back to 4.4 billion years ago (Wilde et al., 2001).

After the Hadean

The birth of plate tectonics, the ultimate recycler

Things started looking up for Earth, and the potential to evolve life on the planet, after four billion years ago when the "Late Heavy Bombardment" of the planet ended. The Archean Eon stretched from 4 to 2.5 billion years ago, and marked the stabilization of many of Earth's geologic and climatic systems, and the evolution of life on the planet. Central to any understanding of how our planet has been so relatively stable for the past 4 billion years is a process just fully recognized by the 1970s–the concept of plate tectonics. It seems arcane to link climate change and evolution to the deep interior workings of our planet, but in fact it is an absolutely required process, on this planet at least. Plate tectonics was effectively "born" toward the end of the Hadean Eon or at the beginning of the Archean Eon (Hopkins et al., 2008) and continues to this day.

The theory of plate tectonics is an outgrowth of the long-ago recognized 3-D jigsaw puzzle that makes up the shape of the continental borders. This idea was put forward, conceptually in the late 1500s and then scientifically by Alfred Wegener in 1912, as "continental drift." Essentially, the east coast of the Americas, for example, fit beautifully along the west coasts of Europe and Africa (assuming that you fit them like edges of a curved orange peel instead of a flat cut edge of a paper). Evidence for this fit was not just geometric when put back together, it linked up fossils and rock types that are now split between two sides of a wide ocean. The concept only gained lukewarm support for lack of a driving force to make entire continents move around on a globe without the planet growing or shrinking. A rapid string of discoveries in the 1960s, followed by observational confirmation in the early 1970s solidified this concept into plate tectonics, where relatively solid and brittle plates on the surface of the Earth are moved around by circulation of a hot, viscous layer below, called the mantle. The surface plates, or lithosphere, are divided between very old and relatively permanent thick continental lithosphere and young, thin oceanic lithosphere. The oceanic lithosphere rides very low on the mantle because it is dense and thin, whereas continental rocks are relatively light and thick.

One clue to the fundamental recycling feature of plate tectonics is that the oceanic lithosphere is uniformly young (by geological standards), with the oldest portions only being 180 million years old and the youngest segments, typically deep underwater in the middle of ocean basins, being actively formed today along long chains of underwater volcanoes. These are segments where new oceanic lithosphere is being born today, effectively shoving preexisting lithosphere to each side of the volcanic segment. These areas constitute the growth of new oceanic lithosphere, essentially adding now surface area to the planet. We have confirmed that the Earth is neither expanding nor contracting, so these regions of oceanic growth must be compensated by areas on the planet where the surface is being destroyed. These segments of destruction are found in areas where older oceanic lithosphere occurs. The oldest oceanic lithosphere is typically found adjoining earthquake regions, which were found to be segments in which oceanic lithosphere is being somewhat reluctantly pushed under continental lithosphere, or in some cases younger and relatively lighter oceanic lithosphere. These are segments of recycling of oceanic lithosphere, back into the viscous mantle where it is remelted.

This is an admittedly simplified version of plate tectonics–generations of scientists have spent their careers studying these processes, as they have major implications not just for understanding how the Earth operates but also for identifying and mitigating natural hazards from plate tectonic-related volcanoes, earthquakes, and tsunamis. But for our purposes, new material emerges from the interior of the Earth along these submarine volcanic chains, called spreading centers, and is plunged back into the deep Earth and recycled at earthquake-heavy subduction zones. Currently, the longest this process takes, from birth to death, is about 180 million years. Subduction largely only occurs with oceanic lithosphere, as it rides lower on the mantle and is thus easier to subduct than thick, light continental lithosphere. This is the reason why we have rocks on the continents that can contain zircon minerals as old as 4.4 billion years the continents are a stable light froth that floats on top of the mantle, occasionally being torn apart as new ocean basins are born and being slammed together when ocean basins shrink to the point of disappearance through subduction.

The recycling component of plate tectonics is really the key to how Earth's climate is stabilized through time, and it is largely through the carbon cycle. All life on Earth is carbon-based, but carbon is also present in many nonorganic forms as well. In the atmosphere, carbon is mainly present as the gasses carbon dioxide, and in lesser abundance methane. These gases are the critical thermostats for our planet, as they are both greenhouse gasses, trapping heat from the sun near the surface. More of them trap more heat, and less trap less. But eventually, those atmospheric forms of carbon get removed by incorporation into plant matter or through chemical reactions related to the weathering of minerals on land or in the sea. Some of that carbon might be returned in a year (think decaying leaf matter in the backyard) or a century or even in millennia, but a portion of it eventually gets caught up in plate tectonics and is subducted into the interior of the planet. In this process, the Earth loses some of its heat-trapping capacity. It happens over millions or tens of millions of years, as the plates on land only move about 5—15 cm per year in any given direction–geology is a slow but inevitable process. Fortunately, the carbon can also be replaced by plate tectonics, through deep sea rifts and volcanoes; otherwise, our planet would forever lose its greenhouse regulation capabilities.

There is actually significant debate about when our plate tectonic engine started. The "standard" model places plate tectonics starting about 3.5 billion years ago, about 1 billion years after the Earth formed, based on modeling and recent work using titanium chemistry in rocks (Plate tectonics may have begun a billion years after Earth's birth, 2017). Other studies, based on pressure and temperature estimates derived from zircon minerals, put that initiation much earlier—indeed, right at the end of the Hadean (A new picture of the early Earth, 2008; Hopkins et al., 2008). An early start to plate tectonics would mean that the recycling component of carbon and water was active before life evolved on Earth, which is consistent with our modern understanding of the requirement of the planet to have plate tectonics in order to have a self-regulating climatic system. But early Earth had temperatures well above freezing because of an extremely strong greenhouse effect, so there was no requirement that carbon be returned to the atmosphere from plate tectonic processes, because there was so much of it in the atmosphere already.

A watery world

Even during the Hadean Era, liquid water was present on the planet, and a lot of it (Earth may have formed with enough water to fill the oceans three times, 2020). The exact origin of that water is still a bit unclear, but the two major contributors were water molecules that were held by minerals and particles that made up the early Earth upon formation and water "adopted" from collisions from asteroids, comets, and other planetary bodies during the turbulent Hadean (New insight into the origin of water on Earth, 2020). Perhaps surprisingly, it is somewhat important that we are able to distinguish the relative contributions of each source, as this provides the framework for understanding the distribution of water in solar systems in general, an important component in our hunt for life beyond our own solar system. Some tools have been used to differentiate the two sources on Earth, including isotopic tracers that are held within the water molecules themselves.

Now, after stating with some confidence that the Hadean was hot and hectic, and that plate tectonics and a full "modern" ocean emerged at the end of the Hadean, somewhat of scientific disclaimer is in order. Much is unknown about this time, and conflicting evidence exists including that plate tectonics emerged during the Hadean, that this Era was actually relatively cool and placid, and that a full liquid ocean theoretically capable of sparking life in its depths. This evidence comes from the same types of records that support the hellish version of the Hadean. This simply speaks to the fact that it is really difficult to reconstruct events from that long ago, from the tiny bits and pieces of geological material left over from that time.

And life emerges ... somehow

The first post-Hadean Era is called the Archean, which stretched from 4 to 2.6 billion years ago. It was during the Archean that life emerged in our solar system, but which planet got life first—Earth, Venus, or Mars? Seems like a strange question to ask, but actually, as life evolved in Earth's oceans toward the beginning of the Archean, it is just as plausible that life would have also evolved in the vast oceans of early Mars or perhaps on Venus. Although we have not explored this planet enough to have direct evidence of Venusian oceans, we have done so on Mars and found that extensive oceans existed in early Mars history. Could life have evolved there? Or, even more provocatively, could life have evolved first on Mars and have been transported to Earth via a meteorite?

During the end of the Hadean and early Archean Eras, asteroid collisions were still frequent enough, and violent enough, to carve out chunks of all of the planets and eject them to outer space, where they could get captured by the gravity of other planets and rain down on them as small meteors. The Earth has a large number of pieces of Mars on it—indeed, bits of Mars rain down on the planet every day. Entire meteorite-hunting expeditions go out to Antarctica in search of Martian meteorites every year, where the actions of glacial movement concentrate rock fragments and meteorites in discrete patches, and the white color of ice makes them easy to see. One Martian meteorite found in Antarctica yielded evidence of intact organic molecules and potential biological structures (Carbon compounds from Mars found inside meteorites, 2020). This finding led NASA researchers to pose the question of whether this was evidence of some kind of simple life on Mars, locked in the interior of this meteorite. The evidence is not convincing enough to support this claim, but it leads to another critical question—could life have evolved on a nearby planet like Mars first, and then have "seeded" our planet via meteorite? This concept is not far-fetched, and is termed "panspermia" by astrobiologists. Evidence from studies on the International Space Station revealed that some bacterium can easily survive a year in space outside of the protection of the vessel itself (This bacterium survived on the outside of the space station for an entire year, 2020), further adding to the credibility that viable microbes can "hitch-hike" between planets on meteors. It thus raises the important question about the origins of life on any planet, and unless genetic testing can be done on actual extant cells to identify the degree of similarity, the evidence that might be retained in fossils does little to help clarify whether life formed uniquely on a particular planet.

The goldilocks concept and planetary habitability

One of the unique aspects of the inner solar system during this time is just how similar the "three sisters" (Venus, Earth, and Mars) were in their surface environments at the beginning of the Archean. Direct and modeling evidence indicates that all three had liquid water on their surfaces, and at least for Earth and Mars, quite a lot of it. They all likely had plate tectonics of some form. They all had dense atmospheres comprised largely of carbon dioxide and devoid of oxygen. And based on direct fossil evidence on Earth, life had evolved on this planet at least. It is unclear how long these conditions persisted on Venus given how difficult imaging is on this planet and how few landing missions have occurred, but various lines of evidence point to the loss of surface water relatively soon into the Archean. And we have direct evidence that these stable conditions persisted for the next 1.5 billion years on Earth while early life was proliferating, if not evolving quickly. But on Mars, more evidence has emerged from chemical analysis of landers, indicating that its watery early history ended after only about 1 billion years.

Why this difference in three planets that started off so much alike? Why did Earth end up being the planet that was "just right" in this Goldilocks analogy? And could Mars and Venus still have life, but in unconventional environments compared to the present Earth? We might not soon, or ever, have the answers to these questions, but clues have emerged from analysis and modeling that the factor behind long-term habitability of these planets is the dynamics of how just one element, carbon, influences planetary temperatures. Too much carbon in the atmosphere makes the planet too hot to retain liquid water (Papa Bear's porridge in this Goldilocks analogy), and too little carbon in the atmosphere makes the planet too cold to retain liquid water (Mama Bear's porridge). Just the right amount (Baby Bear's porridge) keeps waters and oceans present, and just the right amount over a long period provides the time to incubate life on the surface and to allow it to evolve over time—eventually getting to Robert Southey, who would write all about this phenomenon in the fairy tale book Goldilocks and the Three Bears!

The "Goldilocks Zone" is a term used by astrobiologists to define the distance that a planet has to be from its star to have liquid water, broadly assumed to be the key ingredient for life. This zone is scaled to the size and heating capacity of the star—hotter stars would skew the goldilocks zone farther from the sun and cooler stars closer. The planets that are seen orbiting around these stars by our modern "planet-hunting" telescopes are judged potentially habitable if their orbit falls within the Goldilocks zone of its solar system. This definition does not account for long-term changes in the strength of a given star, or whether the surface of the planet has a heat-trapping atmosphere, but hey—when you can even see a planet and define its orbit from 60 trillion miles away—you are doing pretty good! Nevertheless, in our solar system, both factors—changes in the strength of the Sun over time and changes in heat-trapping atmospheres of Venus, Earth, and Mars over time—combined to place us on very climate different trajectories billions of years ago.

The carbon cycle and water

Climate dynamics are covered in great detail in Chapter 2, but briefly for the purposes of comparative planetology, it largely comes down to carbon. On Earth, carbon is balanced between carbon as a gas in the atmosphere (largely carbon dioxide and to a lesser extent methane), carbon on the surface of the planet in soils, plants, and the ocean, and the vast amounts of carbon stored in the deep Earth as limestone, coal, and oil. Modern anthropogenic climate change is largely due to the burning of carbon in coal, oil, and gas and the conversion of this carbon to carbon dioxide, where it acts as a greenhouse gas that warms the planetary surface. Humans have effectively skewed the balance by burning carbon that slowly accumulated over millions of years in geologic deposits in the deep Earth. But on long time scales, this balance is always changing—the key though is that when this balance changes, Earth's climate typically responds.

As discussed at the beginning of this chapter, a unique aspect of this planet is that its carbon thermostat has, for many billions of years, been set between the freezing point and boiling point of water. In Earth's early history, the Sun was dimmer and gave off about 25% less heat energy. This should have resulted in a surface temperature well below freezing for water, pushing the planet outside the effective Goldilocks zone. But this was offset by the Earth having a tremendous amount of its available carbon stored in the atmosphere, which heated the planet and resulted in much of the planet's water present as liquid oceans on the surface. As the Sun slowly warmed and the Earth internal process of plate tectonics matured, the atmosphere began losing some of its carbon to absorption by simple organisms in the ocean and the continual weathering of rocks on the surface. These offsetting factors continue to today. There have certainly been large swings in the amount of carbon in the atmosphere as this "balance of power" between carbon in the deep earth and carbon in the atmosphere varied. Some other factors also influence the surface temperature of the planet, such as reflective ice, but these will be covered in much greater detail later. The key takeaway is that the Earth's carbon thermostat has functioned effectively for billions of years. This was not the case for unlucky Venus and Mars, but for very different reasons.

Venus and a runaway greenhouse

Venus was in the perfect orbital position in the early days of the solar system, being close enough to the dim early Sun to be well placed in the Goldilocks zone. Additionally, it is large enough to have also maintained plate tectonics, which is critical to keep carbon cycling into, and back out of, the planetary interior. But looking at Venus today, you see an extremely dry planet with a choking thick atmosphere of carbon and other gasses, a surface temperature that can reach well above 400°C, and no apparent life. If we could survive on the surface, the thick atmosphere would impart a pressure equivalent to if we were 3000 feet underwater, crushing our bodies immediately. The few Soviet-era landers that ever made it to the surface of Venus snapped photos and sent them back to Earth revealed a rocky lifeless surface (Fig. 1.2). The landers themselves only survived a matter of minutes before the heat and toxic atmosphere caused them to fail. So what went wrong with Venus, and when?

The "what went wrong part" of this is likely relatively straightforward. Venus was close enough to the Sun to begin feeling the early effects of increasing solar output, and thus, its orbital distance was beginning to place it more near the edge of the Goldilocks zone. But what was more critical was that the proximity to the Sun meant the solar winds were more intense at the top of Venus' atmosphere, than is the case for Earth. Because of the early

Figure 1.2 ***Surface of Venus.*** One of the few photographs of the surface of Venus, taken by the Soviet probe Venera 13 in 1982, revealing a rocky, apparently lifeless planet. One study did find evidence of the gas phosphine in the atmosphere, which may indicate the presence of life floating in the smoggy Venusian atmosphere. This, of course, needs to be taken with a huge grain of salt, but is provocative, to be sure. *From* What lies beneath the clouds of planet venus*? (2020). https://www.spaceexploration92.com/2020/02/what-lies-beneath-the-clouds-of-planet-venus.html.*

thick atmosphere like on Earth (but including the additional factor of proximity to the Sun), the atmosphere likely held a significant amount of water on Venus as water vapor throughout the atmosphere, as opposed to Earth, where nearly all of the water is on the surface and what water vapor there is in the lower atmosphere. Carbon is cycled out of the atmosphere via the weathering of rocks and the transport of carbon to the oceans. There, through processes of plant absorption and mineral formation, the carbon deposits to the seafloor and is eventually removed to the deep interior by plate tectonics. In this way, oceans provide the return flow of carbon from the atmosphere to the deep interior. But on Venus, a combination of high solar winds and a "wetter" atmosphere caused Venus water to be literally blown away and lost forever from the planet. As the water vapor was being lost from the atmosphere, more water would be evaporated from Venusian oceans to replace it, just to once again be stripped away by solar winds—a so-called "runaway greenhouse" effect (The runaway greenhouse: a history of water on Venus, 1969). Eventually, as this process continued, the planet lost its rainfall and oceans and their capability to remove carbon from the atmosphere. At this same time, however, carbon continued being brought back to the atmosphere from the inner planet through plate tectonics, but there was no mechanism to remove it again. At some time, a point of no return was reached, and Venus became the hot, lifeless, Papa Bear porridge kind of planet that we see today.

Mars and its deep freeze

Whereas Venus literally had its water blown off into outer space, Mars saw most of its water freeze down onto the surface and into its interior at some point in its middle history. With Mars, the problem was its carbon circulatory system, in a sense. For a planet with the highest volcanic peak of any in the solar system, Mars has very sluggish planet tectonics. Why do we know that? Largely because of its weak magnetic field and thus weak ionosphere. As noted for Earth, a planet with active plate tectonics and a metallic core generates a magnetic field, which in turn generates a protective ionosphere. Also, we now have a seismometer on the surface of Mars (InSight mission overview, 2019). This little lander is a mini-field geologist, fully equipped with a seismometer to measure "Mars Quakes," a drill to measure internal temperature of the planet near its surface and a weather station (to check out the temperature on Mars right now, go to https://mars.nasa.gov/insight/weather/). Dubbed the "InSight," this stationary lander touched down in a crater nicknamed "Homestead hollow" in a region called Elysium Planitia on Mars in 2018 and quickly began reporting the weather and listening for seismic activity. The weather is, as expected, chilly, with daytime highs a not too-unearthly 20°F (C) to a downright chilly −150°F (C). But more importantly, the seismic activity is extremely low, with only a few Mars quakes per week, and all of them weak compared to Earth's prolific quake machine (A year of surprising science from NASA's inSight Mars mission, 2020).

The low seismic activity on Mars indicates that it has little in the way of conventional plate tectonic activity at the moment, and indeed for much of its history. The enormous volcanoes are likely simply from the interior episodically releasing built-up internal heat in the form of eruptions, and doing so all at once in massive single eruptive events rather than continuous smaller eruptions along plate tectonic boundaries, as we see on Earth. Recall that one of the critical recycling mechanisms for carbon on Earth is through carbonate sediments that form in the ocean and rain down onto the seafloor, ultimately to be transported to the deep Earth by subduction and returned to the surface by the continuous eruptions at mid ocean volcanoes. Mars likely quickly lost this recycling mechanism, and with it, the ability for carbon to get back into the atmosphere once it settled to the surface of the planet.

It is unclear when plate tectonics on Mars slowed and then effectively stopped, although the history of its surface water might provide a clue. Mars started out with much less water than the Earth, and theoretically Venus, because of its position in the solar system. There seems definitively to have been a continuous presence of surface water on Mars for its first 700–1000 million years. Even so, the total amount of water on the surface would have accounted for only an ocean that covered about 20% of the surface to a depth of about 140 m. Contrast those figures with the modern Earth, which has an ocean covering about 70% of the planet's surface to an average depth of 3500 m. But even that relatively small amount of water did not last long on Mars. The total amount of water on Mars, in all forms, was 6.5 times higher at the beginning of Martian history (like Earth, Mars, and all the other planets are 4.5 billion years old) (Lost in space: How mars atmosphere evaporated away, 2017). That amount rapidly dropped down to 2 times present after 500 million years, and then has steadily dropped to its current total today (Rickman et al., 2019).

So where did that water go (NASA's MAVEN spacecraft revealed how mars lost its atmosphere and water, 2020)? Some of Mars surface water is still there, but now in the subsurface of the planet in briny aquifers and at the poles as ice. These water sources may occasionally rise to the surface, in the case of the aquifers, or partially melt at the peripheries of the polar ice caps, but it does not remain liquid for long given the freezing temperatures and arid, thin atmosphere, creating seep-like features that are observed by orbiting satellites. But like Venus, most of Mars water was stripped away from the atmosphere. Unlike Venus, however, this atmospheric water loss it was not due to a runaway greenhouse effect, but rather to its loss of plate tectonics. Recall, that plate tectonics are driven by a circulating metal-rich core, and when active, produces a strong magnetic and ionic field that acts as a shield for the atmosphere, protecting it from loss due to solar winds. Mars lost that shield sometime in its early history, and as a result, any water vapor that might enter the atmosphere through sublimation or evaporation is at risk of being lost from the planet, forever.

It is difficult to determine when the last large bodies of liquid surface water existed on Mars, but we know one thing for certain–several of the large impact craters in the equatorial region of Mars held lakes of water, some as "recently" as three billion years ago. How do we know that? Because we have sent a water detective to one of them! The Curiosity rover landed in and explored Gale Crater, which features layers of sediment deposited by water, and has clay minerals that only form when liquid water is present.

Curiosity specifically went to Gale to explore its watery history, as it is similar to the age and size of a number of other impact basins that appear to have hosted lakes. How long the lake in Gale Crater lasted is still an open question, but the clear signs of water flow (Fig. 1.3), with images that easily could have been taken on Earth (Fig. 1.4), show environments on the surface that had similar conditions to those of early Earth, and thus very well might have been the birthplace of Martian life. Curiosity also unearthed evidence of massive flooding events, perhaps the result of impacts on a then mostly ice-covered surface which released heat and triggered rapid melting of surface ice (Curiosity finds evidence of ancient megaflood on Mars, 2020).

Plate tectonics on Mars eventually slowed to a climatic tipping point, such that its carbon release back to the atmosphere was simply too slow to return enough warming carbon dioxide to the surface to keep temperatures consistently above freezing. With that tipping point reached, all liquid surface water would have frozen onto the subsurface, and water in the form of water vapor rapidly lost from the atmosphere, further decreasing the planet's greenhouse effect and setting the great Mars freeze in place for good. But was that the last time that liquid water appeared at the surface? Multiple lines of evidence indicate that Mars indeed has had liquid water

Figure 1.3 ***Sedimentary Rock Layers on the Surface of Mars.*** Sedimentary rock layers in Gale Crater on the surface of Mars, indicating the presence of liquid water about 3 billion years ago. *This photograph was from NASA's Curiosity rover (https://lavinya.net/resim/363/marstan-bir-gorunum).*

Figure 1.4 ***Sedimentary Rocks on the Surface of Mars.*** A photograph taken by Curiosity rover in Gale Crater, showing an outcrop with finely layered rocks on Mars.

at the surface now and again. High-resolution satellite imagery of Mars has shown multiple sites with contemporary liquid water flowing on the surface, likely in the form of a water-sediment slurry flowing down hillslopes (NASA confirms evidence that liquid water flows on today's Mars, 2015) (Fig. 1.5). This is not particularly surprising, because the equatorial regions on Mars do

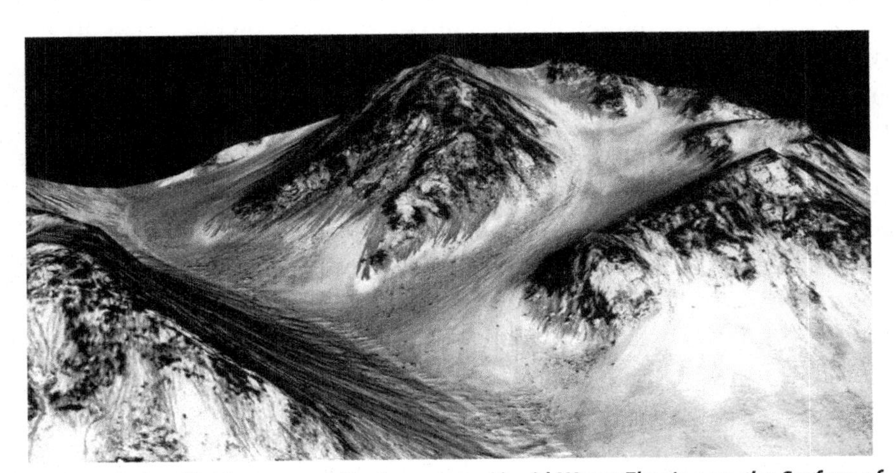

Figure 1.5 ***Satellite Imagery of Contemporary Liquid Water Flowing on the Surface of Mars.*** Dark, narrow streaks on Martian slopes such as these at Hale Crater are inferred to be formed by seasonal flow of water on contemporary Mars. The streaks are roughly the length of a football field. *Image credit: NASA/JPL-Caltech/Univ. of Arizona.*

get above freezing at times, which due to the thin atmosphere on Mars is actually −93°C (−136°F) (Nair & Unnikrishnan, 2020). But large volcanoes active as recently as 2−115 million years ago (Recent activity on Mars: Fire and ice, 2005) (we know their young age because they are relatively unblemished by craters from meteorites) might have provided local heat sources for melting, at least in the subsurface. And a 2020 finding reveals vast lakes of liquid water under the surface, likely heated by recent volcanic activity and begging the question of whether life might persist on Mars, but at depth (Buried lakes of liquid water discovered on Mars, 2020). This is certainly the case for Earth, which has a vast subsurface ecosystem, dominated largely by microorganisms that can no longer tolerate the oxygen-rich conditions on the surface (Buried lakes of liquid water discovered on Mars, 2020).

The long Archean, and even longer Proterozoic

The Archean Eon and the formation of life

The Archean Eon, stretching from 4 to 2.5 billion years ago, is followed by an even longer one, the Proterozoic Era which stretches from 2.5 to 0.54 billion years ago. These eons are long because historically they left very little record of themselves in the rock record, and there really were not any obvious extinction events, one characteristic that defines the divisions of the geologic record. The "boring" does not show up in the geologic record, but certainly dominates Earth history. More recent work reveals that much more went on during these eons, and many of these events were literally Earth-transforming. Indeed, much of this book focuses on these transformative intervals because they reveal much about how the Earth functions as a system, and how coupled climate−ecosystem dynamics have played out over time, and continue to today.

Sometime during the Archean Eon, the right processes and ingredients (Underwater discovery prompts shakeup of long-held theories for origins of life, 2020) came together to form the first life on this planet. Much about these processes and ingredients Chapter 3 ultimately remain in shadow, obscured by the veil of time. But based on myriad lines of evidence (Donoghue, 2020), including genetic models (Did life emerge in the "primordial Soup" via DNA or RNA? maybe both, 2020) and molecular theory biology (Javaux, 2019), as well as scant fossil evidence, among the first group of organisms that appeared on Earth were related to Archaea. These single-celled organisms lack a nucleus inside of the cell and are quite simplistic in metabolism and structure. Archaea or their ancestors (Koch, 1998) were likely among the first of the organisms that evolved that persist to today (New insights into the evolution of complex life on Earth, 2020). One of them, which produces methane as part of its metabolism, can be found in modern environments that are devoid of oxygen, which is toxic to them. But myriad other early organisms, such as bacteria, also persist to this day, and their metabolisms have driven many global biological and chemical cycles. As covered in Chapter 4, the most critical metabolic byproduct of early organisms is the gas oxygen. This is not just critical to humans and all the other animals that have ever lived on the planet, all of whom require free oxygen for life, but also for the Earth's geologic development and its climate. Archaea do not produce oxygen, but other early organisms did. But for this brief tour of deep time, the key takeaway during the Archean Era is that life evolved, likely in a number of forms, and a class of them began producing oxygen, and lots of it (Did life emerge in the "primordial Soup" via DNA or RNA? Maybe both, 2020).

The Proterozoic eon and the emergence of an oxygen world

The beginning of the Proterozoic is marked by the emergence of free oxygen on the planet. Microbes had been kicking oxygen out of their cells for hundreds of millions of years, but this was all consumed by geochemical reactions involving iron at such a great rate that none of it accumulated as a gas, either dissolved in ocean water or in the atmosphere. Finally, the geologic sink for oxygen started to become saturated, and it began bubbling into water and eventually the atmosphere. Note that this process of biological oxygen formation occurred exclusively in the water—for reasons covered in detail in Chapter 3, life needed a protective ozone shield to survive above the water, but water itself absorbs the harmful high-energy radiation from the sun. The accumulation of free oxygen, during an interval at the beginning of the Proterozoic called the Great Oxidation Event, literally transformed the surface of the planet, and led to the evolution of more complex organisms including animals.

Our geologic record of the Proterozoic is much more complete than for the earlier Archean, and thus we are able to define a number of critical events that occurred during this long eon. In brief, because these events are all covered elsewhere in this book and in much great details, the Proterozoic started with the Great Oxidation Event (Gaucher & Frei, 2018), which was followed closely by the planet's first near-global glaciation event, underwent various changes in oxygen levels and related levels of nutrients and carbon, saw the first beginnings of a stratospheric ozone layer and the evolution of animals, and closed with several additional near-global glaciation events that might have ultimately spurred the most dramatic interval of evolution that the planet has ever seen. One of the reasons why several other chapters in this book are devoted to events during the Proterozoic is that scientific findings in this eon have exploded over the past several decades. Even an interval from 1.8 to 0.8 billion years ago in the middle of the Proterozoic that used to be called the "Boring Billion" because nothing seemed to have been happening might actually be one of the most influential times in Earth's history.

The Proterozoic also marks the beginning of our ability to track the movements of continents using geomagnetic signatures in rocks. Our magnetosphere has been present since the birth of plate tectonics and two characteristics of this magnetosphere make it critical in deciphering the latitude and migration patterns of rocks. First, the magnetic fields flip poles episodically (this is covered in more detail in Chapter 2) so that the magnetic "north" pole flips to the south pole. The patterns of flips over time become a time marker, much like the thickness patterns of tree rings can be used to reconstruct the archaeological past. Second, the magnetic field is practically horizontal near the equator, but dips in sharply where the field emerges and descends at each pole (Fig. 1.6). The angle of this field is similarly trapped in the mineral matrix of rocks and sediments formed at that time, becoming a

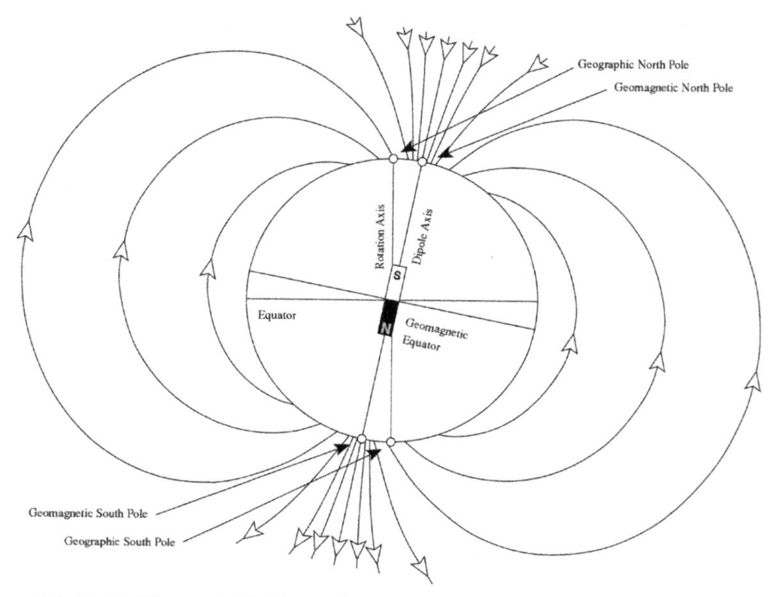

Figure 1.6 ***Earth Magnetic Field.*** An illustration of the tilted dipole approximation to the Earth's magnetic field, showing the horizontal flow patterns near the equator and the sharply dipping ones near the poles that can be used to track the movement of continents.

latitude marker. Collectively, these signatures reveal the earliest known supercontinent, Rodinia, which began breaking apart about 750 million years ago and is comprised of various continental plate fragments that have recombined and again broken apart since.

The Phanerozoic and the emergence of our modern world

The most recent period of geologic history, the Phanerozoic, began with an evolutionary explosion and ends with us—human beings. The geologic record becomes more and more complete as the clock of Earth time ticks toward the present, and thus, the interplay between Earth systems and life becomes a bit simpler to track. This is not to say that the science behind these connections is settled, and indeed major hypotheses about earth systems in this period are forwarded and debated in conferences and journals every day. Some of these are minor tussles about the naming of species, or the exact timing of various events. But some are still profound unknowns that deeply influence our own future. For example, we know that the levels of greenhouse gases in the atmosphere varied widely during the Phanerozoic and strongly influence global temperatures and ecosystems. But we do not yet know certainly how much extra heating we get for each additional amount of carbon dioxide in the atmosphere, thus providing one of the uncertainties with projecting how human activities will warm the planet. But we certainly know from the geologic record that the "modern" ecosystems of the Phanerozoic were influenced heavily by the course of evolution, by climate change, and by a number of profound events, such as global extinction events that led to the demise of large swaths of organisms on land and in the sea.

Of the modern systems that formed during the Phanerozoic, Chapter 5 describes in detail how plants evolved roots by making a new, tougher biomolecule around 370 million years ago. This allowed plants to "mine" the nutrients present in rock, and in so doing created the world's first soils. This development permanently reshaped the surface interface between rock and atmosphere, profoundly impacting global cycles of nutrients and carbon, and changing climate. These rooted plants became the first forests, again owing to the tougher biomolecules that plants formed in a sort of arms race to be the tallest plants around to capture sunlight and outcompete others.

The Phanerozoic saw multiple massive extinction events that reshuffled ecosystems and reset various biological lineages. As life evolved to become more complex, it also diversified its abilities to fill ecological niches, and to refill niches when the current occupants were wiped out by extinction events. These processes are detailed in Chapter 8. Although new details are emerging all the time about the characters and sharpness of these extinction events, they seem mostly related to either geologic events, such as massive volcanism, or extraterrestrial ones, such as the massive asteroid that killed off the nonavian dinosaurs about 65 million years ago. It is important to recognize, however, that it was not necessarily the eruptions or the collisions that drove the severity of these extinction events, but rather the profound influence that they had on global climate, which then drove extinctions. Thus, extinction events are the most obvious example of climate change and life interactions that are detailed throughout this book. Additionally, understanding the scope and scale of these past mass extinctions allows us to place some perspective on our current, so-called "sixth mass extinction." This extinction event is again largely related to climate change, but not as a result of geological or extraterrestrial catastrophes, but rather to one species only—*Homo sapiens*. This will be covered in great detail in Chapter 9.

Finally, the end of the Phanerozoic marks the most recent of several global transitions from a warmer planet, so-called Hothouse conditions, to a colder planet, so-called Icehouse conditions. The transition from a Hothouse Earth about 50 million years ago through several steps into our current Coldhouse conditions, marked by periodic and massive ice ages, is illustrative of how Earth systems interact with each other in relatively complicated ways, including the shifting of plates and the birth and erosion of mountains, to influence the balance of carbon on the planet (covered in detail in Chapter 2) and thus global climate. It is on this template of recent climate swings that hominids evolved, and indeed their persistence, through to the present incarnation in us, may in fact be linked to their unique capacity to adapt to climate change not restricted simply to migration as a solution but also fire and clothing—the first organisms to intentionally utilize either of these mechanisms for survival.

Summary

From fire to ice, so is the long history of the Earth's climate. The major episodes of Earth can seem like a recitation of unconnected events, but in fact, they are the result of dynamic systems that have dictated Earth's heat balance and defined the birth and progression of life on the planet. It is a bit stunning to think that, only a few hundreds of millions of years since its very fiery and chaotic creation, life was formed on the planet. Even more amazing perhaps is that the early Earth was none too gentle, marked by continued bombardment from meteors as the solar system was cleared by its planetary vacuums the planets and an atmosphere that would be poisonous to life as we know it. And yet life clung on. Although it is difficult to track too many details about early Earth climate, we at least know that the oceans were persistent through this early chaos and thus were a safe home for these early ecosystems.

In its middle age now, the Earth has revealed amazing resilience as it went through several global shocks, including an entire change in its global biogeochemical composition with the build-up of oxygen, several global ice events, life creeping up on land and taking residence, and now humans completely reshaping global climate and ecosystems in a span of a handful of generations. Through it all, Earth's thermostat remained at a remarkably steady 0—100°C, a feat its sister planets Venus and Mars could not match. It is interesting to conjecture whether this same arc of long climate stability is a necessary requirement for the evolution of intelligent life. And if so, how many other planets in the vast cosmos have achieved this stability? Further, on our planet it took about four billion years for life to cook from algae to humans—are we particularly slow in universal timescales? Would more instability in Earth's climate have sped evolution, as organisms would have had to adapt and evolve more frequently? Certainly in Earth's history the intervals when climate conditions changed most rapidly and dramatically coincide with more extinction, and radiation, of organisms. And hominids and we humans evolved in a time of deeply chaotic swings from hot to cold, and wet to dry. Perhaps even human-caused climate change might spur evolutionary events, as it certainly has spurred a severe mass extinction. For now, though, we can be happy to be inhabitants of a livable planet, 4.5 billion years old and counting.

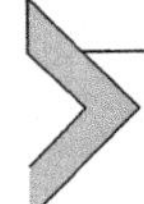

Did you know that?

The same scientist who provided the first accurate date of the earth also is behind why two major environmental laws were passed in the USA

Dating the Earth was really hard, and the fact that early geologists worked largely from a descriptive lens without fully recognizing that the layers and structures that are observed in the rock were formed in processes similar to those that were occurring all around them did not help. Some attempted to use scripture as the time keeper. The most well known of these was Bishop Ussher of Ireland. Ussher based his chronometer on a literal reading of the Old Testament. When this bible listed the "Jonas begat Ezekiel, who begat Abel, who begat ...," Bishop Ussher counted all of these as generations, with each generation spanning a certain amount of time. He got quite specific with age estimate, proposing that God began creating the universe (and Earth) on October 23, 4004 BCE (note that he did not specify a time, which seems a major shortcoming of the work ...). His 1650 book, with the short title of "Annals of the Old Testament, deduced from the first origins of the world, the chronicle of Asiatic and Egyptian matters together produced from the beginning of historical time up to the beginnings of Maccabees," remained a piece of religious and scientific dogma for almost 200 years before falling out of favor.

During the 20th century, however, geoscientists started taking full advantage of the constant decay rate of various radioactive elements found in various geologic materials to tell time. Carbon-14 dating proved critical for dating various materials in geologic strata in the recent past, as it is a relatively reliable chronometer back to about 200,000 years ago. Because C-14 is incorporated into plant matter when it grows, it is the plant material itself, or various products of plant matter like charcoal, that is used for the dating and has proven critical for understanding anthropological and archaeological reconstructions of human activity. The full age of the Earth, however, was first accurately determined in Pasadena, California, on the campus of the California Institute of Technology. There, the geochemist Claire Patterson had set up a lab to measure the content of various heavy isotopic elements, such as lead and uranium, that decay so slowly that there are still remnants of the radioactive version of the elements in geologic material that is billions of years old. This was challenging work, and to avoid contamination by outside sources, he had to build a first-of-its-kind "clean laboratory" to do the various rock extractions and house the mass spectrometer that actually measured the elements.

In 1956, Claire ultimately successfully and accurately determined the age of the Earth (4.55 billion years) using lead isotopes, but in the process also discovered that the contamination of materials from the element lead was everywhere. He overcame those contamination issues to date the Earth, but then, in exploring where the lead was coming from, he found that the culprit was us. Lead emissions from the combustion of leaded gasoline in the car-loving Los Angeles area were turning up everywhere, and not just on lab benches but in people, where it does irreparable damage, particular to the brains of children. Claire then went on an environmental crusade to fight the companies that were adding lead to products like gasoline, paint, and water pipes, facing considerable resistance from corporate forces before finally producing one of the greatest environmental victories of the 20th century in the United States of America—the passing of the Clean Air and Clean Water Acts that eliminated sources of lead and other harmful materials to the environment. Rarely has one scientist had such astounding successes on the two ends of the spectrum of Earth's history.

References

A new picture of the early Earth. 2008. https://www.nytimes.com/2008/12/02/science/02eart.html?8dpc.

Buried lakes of liquid water discovered on Mars. 2020. https://www.bbc.com/news/science-environment-54337779.

Carbon compounds from Mars found inside meteorites. 2020. https://www.nasa.gov/mission_pages/mars/multimedia/pia00289.html.

Curiosity finds evidence of ancient megaflood on Mars. 2020. https://newatlas.com/space/ancient-megaflood-water-mars-curiosity/.

Did life emerge in the "primordial soup" via DNA or RNA? Maybe both. 2020. https://phys.org/news/2020-06-life-emerge-primordial-soup-dna.html.

Donoghue P. Fossil cells. Current Biology 2020;30(10):485–90. https://doi.org/10.1016/j.cub.2020.02.063.

Earth may have formed with enough water to fill the oceans three times. 2020. https://www.newscientist.com/article/2252735-earth-may-have-formed-with-enough-water-to-fill-the-oceans-three-times/.

Gaucher C, Frei R. The Archean-Proterozoic boundary and the great oxidation event. In: Sial A, Gaucher C, Ramkumar M, Ferreira V, editors. Geophysical Monograph Series. American Geophysical Union; 2018. https://doi.org/10.1002/9781119382508.ch3.

Hopkins M, Harrison M, Manning C. Low heat flow inferred from >4 Gyr zircons suggests Hadean plate boundary interactions. Nature 2008;456:493–6. https://doi.org/10.1038/nature07465.

InSight mission overview. 2019. https://mars.nasa.gov/insight/mission/overview/.

Javaux E. Challenges in evidencing the earliest traces of life. Nature 2019;572:451–60. https://doi.org/10.1038/s41586-019-1436-4.

Koch A. How did bacteria come to be? Advances in Microbial Physiology 1998;40:353–99. https://doi.org/10.1016/S0065-2911(08)60135-6.

Lost in space: How Mars' atmosphere evaporated away. 2017. https://www.space.com/36277-mars-thick-atmosphere-lost-in-space.html.

Nair R, Unnikrishnan V. Stability of the liquid water Phase on mars: a thermodynamic analysis considering martian atmospheric conditions and perchlorate brine solutions. ACS Omega 2020;5(16):9391–7. https://doi.org/10.1021/acsomega.0c00444.

NASA confirms evidence that liquid water flows on today's Mars. 2015. https://www.nasa.gov/press-release/nasa-confirms-evidence-that-liquid-water-flows-on-today-s-mars.

NASA's MAVEN Spacecraft revealed how Mars lost its atmosphere and water. 2020. https://jatan.space/nasa-maven-mars-orbiter/#:~:text=Without%20much%20of%20an%20atmosphere,the%20presence%20of%20liquid%20water.

New insight into the origin of water on Earth. 2020. https://www.sciencedaily.com/releases/2020/07/200717120158.htm.

New insights into the evolution of complex life on Earth. 2020. https://New-insight-into-the-evolution-of-complex-life-on-Earth.

Plate tectonics may have begun a billion years after Earth's birth. 2017. https://www.livescience.com/60478-plate-tectonics-gets-new-age.html#:~:text=No%20one%20has%20ever%20been,the%20formation%20of%20the%20planet.

Recent activity on Mars: fire and ice. 2005. http://www.psrd.hawaii.edu/Jan05/MarsRecently.html.

Rickman H, Błęck MI, Gurgurewicz J, Jorgensen UG, Słabyd E, Szutowicza S, Zalewska N. Water in the history of Mars: an assessment. Planetary and Space Science 2019;166: 70–89. https://doi.org/10.1016/j.pss.2018.08.003.

The runaway greenhouse: a history of water on Venus. Journal of Atmospheric Science 1969;26(6):1191–8. https://doi.org/10.1175/1520-0469(1969)026<1191:TRGAHO>2.0.CO;2.

This bacterium survived on the outside of the space station for an entire year. 2020. https://www.livescience.com/bacterium-survives-year-on-space-station.html?utm_source=notification.

Underwater discovery prompts shakeup of long-held theories for origins of life. 2020. https://www.ucalgary.ca/news/underwater-discovery-prompts-shakeup-long-held-theories-origins-life.

Wilde S, Valley J, Peck W, Graham C. Evidence from detrital zircons for the existence of continental crust and oceans on the Earth 4.4 Gyr ago. Nature 2001;572:451–60. https://doi.org/10.1038/35051550.

A year of surprising science from NASA's insight Mars mission. 2020. https://mars.nasa.gov/news/8613/a-year-of-surprising-science-from-nasas-insight-mars-mission/?site=insight.

CHAPTER TWO

Climate drivers on Earth

One of the most frequent rebuttals that climate change experts hear to the facts behind anthropogenic global warming (AGW) is "but isn't climate always changing?" Although on rare occasion this is meant as a real question, it is really meant to undermine two basic premises of AGW. First, because climate is always changing, as geologists will confirm for Earth's entire history, how do we know that current climate change is not just "natural" variability? Second, that even if human-caused, it really is not a big worry as it has happened in the past, and will do so in the future. This chapter will provide tools not only to understand the partial validity of both of these statements but also to reveal that they are both wrong as a counterargument to AGW. Climate is driven by a number of factors and has varied well before humans came on the scene and will do so long after we are gone, with wild swings toward warmer and colder, states. But never has climate changed this much and this quickly and thus, it is critical to put AGW in the context of geologic time by examining the various drivers of climate.

Broadly, we can consider two kinds of climate drivers to the Earth system—one external to the planet and one intrinsic to the planet. Differentiating these drivers has been a challenge to scientists for centuries, as geologists and naturalists began to uncover evidence that some places on Earth, presently in cold regions, sustained tropical rainforests in the past. Similarly, rock records in tropical regions revealed clear evidence of glaciation and ice sheet movement. As we started to gain a better understanding of Earth's past climate through geologic records, we started to notice patterns in time and space, and began to leverage basic principles of physics, chemistry and biology to expand on frankly rather sparse geologic evidence to build a more coherent model of the Earth's climate system. Think of this as the recipe book of climate—all ingredients can be varied and in variation, a baker can expect different outcomes.

On a simple level, Earth's climate at any one time is a balance between how much energy is received by our planet and how much escapes. The energy that remains essentially dictates global climate. The amount of energy received is the external driver of climate, and the amount retained is largely driven by intrinsic processes on Earth. Both of these factors vary and it is only

Climate Change and Life
ISBN: 978-0-12-822568-4
https://doi.org/10.1016/B978-0-12-822568-4.00005-5

in the last century that we have pieced together how much they relatively contribute to Earth's net climate. But at this point, our Earth System Climate model is sophisticated enough to both explain much of the past variation, and using that knowledge to predict future variation under a variety of scenarios.

External drivers of climate

Earth's only substantial energy source is radiation from the Sun, an enormous fusion reactor only 93 million miles away. Sure, we have residual heat still inside the planet, retained from our original fiery formation. And we are bombarded by extraplanetary particles all the time, most extremely small, and a thankfully rare few that are large—called "civilization ending impactors." These bombardments impart their own energy to our planet. Both the residual heat and the bombardment components of Earth's energy budget are quite small compared to the enormous amount of energy that we receive from our star.

Two major factors influence the amount of solar energy that Earth receives over time. The first factor is fascinating but rather trivial as energy goes—every 11 years or so, the Sun goes through a cycle of more energy output, the less output. We can see this even looking at the face of the Sun, as it correlates to the number of sunspots there are. During low solar output, there are no sunspots, and during high output, there are lots of sunspots. This is caused by internal pulsation of the Sun's fusion reaction core, like the Sun's heartbeat. The variation in energy that we receive from the Sun due to this factor is very small (less than 0.1% of the total energy balance) and thus, although fascinating and easily observable by tracking sunspots, it is simply not a big player in Earth's climate system. The Sun does occasionally throw out massive flares of extrasolar material and radiation energy, substantial enough to cause disruption in the Earth's magnetic field. But these flare events are also relatively inconsequential as net energy drivers for the Earth.

The most impactful external driver of climate on Earth is our very own orbit. People in the northern latitudes see this driver written very clearly on the landscape of low rounded hills, vast flat plains, and striated rocks. These landforms all bear evidence of the power of the orbit to cause Ice Ages, with massive ice sheets and glaciers plowing through and transforming everything in their path. The most recent one only began to let loose its icy grip 20,000 years ago, yielding the so-called "Interglacial" period that we are living in today. Indeed, our *Homo sapiens* species have lived through two Ice Ages, and our interbred cousins, the now-extinct Neanderthals, a few more before that.

One of the first clues to the mechanism of this orbital driver was the observation that the Earth has not experienced 1, 2, or even 10 Ice Ages, but rather countless ones stretching through billions of years of Earth history. Some were massive, shrouding the planet almost completely in ice, some were subtle, barely showing up in the geological record, but most were quite regularly paced, providing evidence for a periodic pattern in energy received by the Earth. But the trigger for all of these was variations in Earth's orbit, which caused variations in how much energy we received from the Sun, where on the planet that energy was received, and when during the year it was received (https://climate.nasa.gov/news/2948/milankovitch-orbital-cycles-and-their-role-in-earths-climate/). The story is a bit more complicated than this (as most stories are!), and later we explore the additive feedbacks between the orbital driver and various intrinsic players on Earth. But for now, let us focus on orbits and energy, and specifically, the energy variations that are regularly paced through time.

If you want to blame anything for our Ice Ages, blame the Moon. The Moon formed very early in the original formation of the Earth, the result of a cataclysmic collision between the newly solid Earth and a planetary body about the size of Mars (Fig. 2.1), called "Theia," a mythological Greek titan who is the daughter of the sky and the Earth and the mother of the sun,

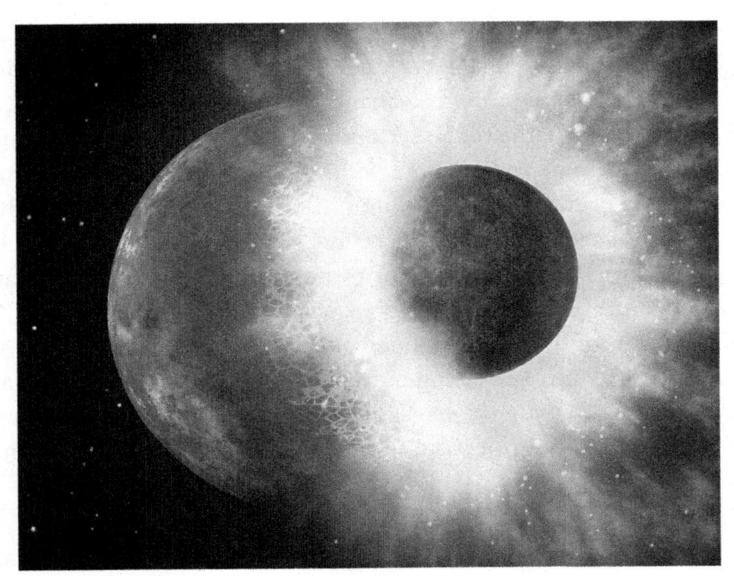

Figure 2.1 ***Theia Collision With Proto-Earth.*** A planetoid roughly the size of Mars collided with the Earth soon after it formed, causing a remelting of the planet and ejecting a mass that soon solidified into the Moon.

moon, and the dawn. The collision was intense enough to temporarily melt the Earth as the collider basically melded with the planet (https://www.space.com/19275-moon-formation.html). In the melting phase, a large portion of the Earth was ejected, trapped by the Earth's gravity, and locked into place. This mass, of course, is the Moon, and it has been with us since the beginning (https://www.extremetech.com/extreme/285173-new-nasa-study-suggests-moon-is-made-of-material-from-earth-not-theia). It is a uniquely large moon compared to the size of the planet, and has a mass and gravity that has powerful impacts on the Earth, ranging from slowing down our orbit over time (our day length used to be closer to 16 h) to being a primary driver of ocean tides.

That early collision caused two of the most critical factors in controlling energy amount and distribution on Earth to this day. First, it altered our spin axis from what was originally vertical to the solar plane to our current rotation angle of about 20% off vertical (although this angle varies through time). The Earth spins around a perpetually tilted axis as it travels its annual orbit around the Sun. This profoundly shapes weather, climate, and ecosystem distribution on our planet as it produces seasons. At the apex of Northern Hemisphere summer (June 21), the North Pole tilts toward the Sun and there is perpetual daylight in the far northern latitudes and perpetual darkness in the far southern latitudes. Six months later, the opposite occurs. The strong seasonality in and of itself has shaped the evolutionary trajectory of life on the planet.

Sticking with the tilt angle a bit longer, the second result of the Moon-forming collision is that it caused a variable wobble in Earth's tilt angle, such that every 41,000 years, the tilt goes from 22.1 to 24.5° from the solar plane, and back again. This periodic wobble in the tilt angle causes a net solar energy distribution pattern called Obliquity (Fig. 2.2), or how obliquely the Earth's surface is facing the Sun. This orbital variable has a subtle but important impact on the distribution of solar energy input. At a high tilt angle, 24.5° from vertical, more energy is received at higher latitudes in both the northern and southern hemispheres at the peak of their respective summers, because these areas are facing more directly toward the sun. And at the low tilt angle of 22.5°, the higher latitudes received much less direct solar energy. The total amount of solar energy received by Earth is the same in either scenario.

A further product of Earth's collision-marred birth is that its orbit is not regularly circular around the Sun. On time periods of about 100,000 years, the orbit goes from being close to circular (Fig. 2.3) to quite oval, and back

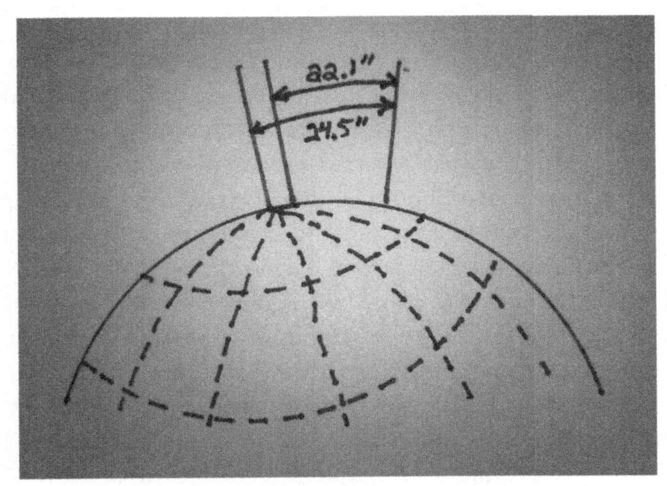

Figure 2.2 ***Obliquity of Earth's Orbit.*** Obliquity is the change in the Earth's tilt with respect to the solar plane, and ranges from 22.1 to 24.5°.

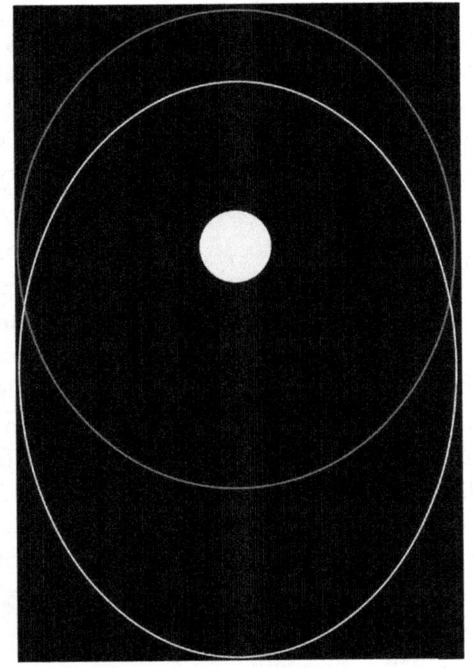

Figure 2.3 ***Earth's Orbital Obliquity.*** Earth's orbit around the Sun varies a nearly circular mode to an oval one with an orbital distance from the Sun that varies significantly over a year.

again. This variable orbital parameter is called eccentricity, and has a substantial impact on the total amount of solar energy that the Earth received, and energy extremes over a given year. That is because in oval mode, the orbit shifts such that the Earth is a bit closer to the Sun at one time of the year, but much further from the Sun than in circular mode just 6 months later. Winter and summer are caused by obliquity as described above, but how cold that winter is, or hot that summer, is influenced by eccentricity. Consider the Northern Hemisphere summer, when the north is tilted toward the sun. When the orbit is near circular, annual solar energy input would be relatively consistent year to year, for tens of thousands of years. But as the orbit slowly shifts to a more oval configuration, the Northern Hemisphere will receive much more solar energy when summer coincides with the narrow part of the oval because the Earth will be closer to the Sun. The inverse holds as well—if instead the Northern Hemisphere coincides with the widest part of the oval, summer will be colder. In a nutshell, a more circular orbit yields more consistent energy inputs, whereas an oval orbit yields long periods of extremes. It is during the extreme periods that the Northern Hemisphere summer coincides with the most distant part of an oval orbit that snow that fell at high latitudes during the preceding winter does not melt all the way. Year after year of snow piling and turning to ice slowly grows glaciers. These growing glaciers coalesce as ice sheets, and as the ice sheets grow thicker and thicker, their weight starts moving them out, like adding sand on a sandpile until the pile cannot get any higher due to gravity and starts growing out.

The final orbital variation that impacts the distribution of energy that the Earth receives is called the Precession of the Equinoxes (Fig. 2.4), or simply Precession. This occurs on time scales of 19–23 thousand years. Those astronomy-savvy readers will know that our rotation axis points to the North Star, so-called, of course, because our North Pole always rotates around it and thus it is the only star in the sky that does not move during an evening, or indeed during the year. But just about 10,000 years ago, our "North Star" would instead be the star Vega.

These various orbital variations were first described in a comprehensive way in the 1920s by Milankovitch, a Serbian geophysicist and astronomer, who also provided the tools to model their impacts back through time. Because these variations are each independent of each other, their net impact on controlling how much solar energy the Earth receives and the distribution of that energy, has to be viewed collectively, as the sum of the impacts.

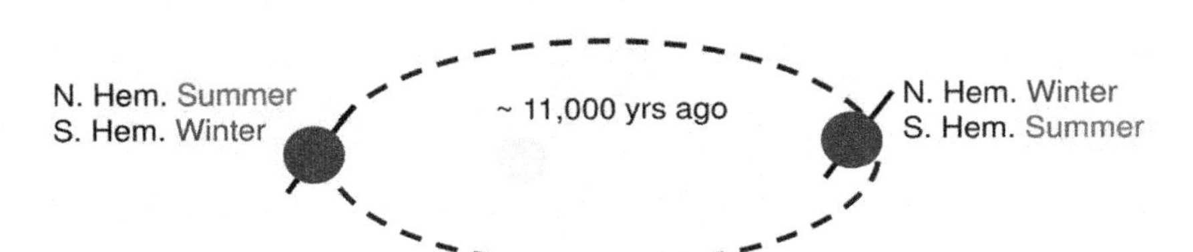

Figure 2.4 ***Precession of the Equinoxes.*** The direction of the Earth's spin axis varies on about 20,000 year cycles, ranging from pointing directly at the North Star like today to directly at the star Vega.

This sum is called solar insolation, and is defined by how much solar energy the Earth receives.

The cycles described by Milankovitch, and the resultant changes in insolation, are not just theoretical, but are recorded in the geologic record. One clear example of this can be seen over the past ~3 million years, a time period in which the Earth has experienced strong, periodic Ice Ages, called glacial periods, interspersed with warmer intervals, called interglacial periods—which we are experiencing now. The periodicity of these events can be determined both in the geologic record and in the record of ice accumulating in the polar regions. Both of these types of materials hold various components that can be age dated, and in the case of accumulated ice, down even to the particular year (Fig. 2.5) when it accumulated.

These past records reveal several interesting aspects of how orbital variations interact with insolation and earth surface heating. But first, it is important to understand why the distribution of solar energy, and not just the amount, is critical. Because of the vagaries of plate tectonics and the current configuration of continents on Earth, the Northern Hemisphere holds most of the land mass, and the Southern Hemisphere the least. Particularly important is that at the latitude of the **Arctic Circle** (Fig. 2.6), nearly the entire

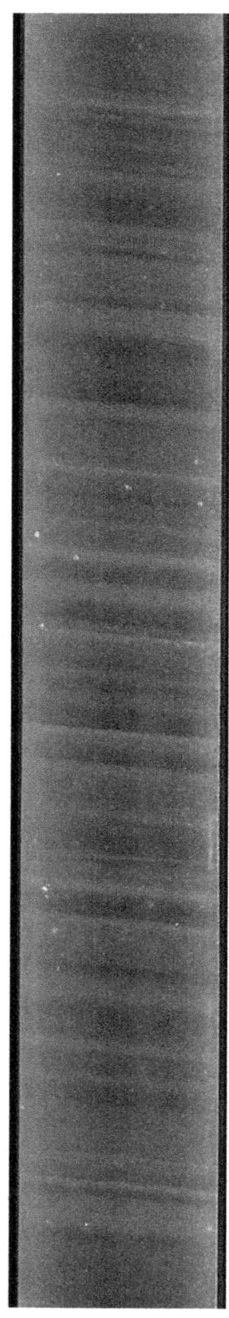

Figure 2.5 ***Ice Core.*** Ice cores reveal the interior structure of an ice sheet and typically record annual layers, like a tree ring, that can be age dated back through time.

Figure 2.6 ***Arctic Circle at 65° North Latitude.*** The Arctic Circle is nearly completely over land, causing it to be sensitive to the amount of solar energy received via the ability to accumulate ice year after year in a low energy phase. The accumulated ice can grow into massive ice sheets, which then travel southward.

Earth surface is land. Why is this important? Because ice sheets can only accumulate on land, and if summers at this latitude were cooler than typical because they received less insolation due to orbital variations, the snow that fell during the preceding winter would not completely melt and would instead begin accumulating year after year. After a few millennia or so, this accumulated ice would have a life of its own, able to grow and migrate southward, beginning the steady growth and advance of the Northern Hemisphere ice sheets that have dominated glacial periods over the past 2.8 million years.

From about 2.8 million years to roughly 800,000 years, the repetitious cycle of glacial periods came every 40,000 years, as reflected well in the geologic record of microfossils on the seafloor. So every 40,000 years was marked by a glacial period, the result of a low in insolation at the Arctic Circle that drove ice accumulation. This time period is consistent with obliquity variations in our orbit being the dominant external factor that controlled global climate. The shorter precession signal can be seen in the geologic record as more minor variations in temperature during this time, riding on the back of the major swings in global climate driven by obliquity. Then, for reasons still being debated, the glacial cycles began to have a 100,000 year repeat beginning around 800,000 thousand years ago, at which point it seems that the dominant driver of glacial cycles on Earth switched to eccentricity in the orbit. Again, during this time period, the precession signal is observed, but is not dominant.

The geologic signal of the precession cycle is present in rock deposits worldwide and helps to determine the timing of various climatic and geological events. Some settings are more sensitive to the precession signal than others. One example of a sensitive record is literally painted across the hillslopes of the Almeria region in southern Spain, manifesting as stripes of different colors (Fig. 2.7) on the arid hillsides. These color differences are due to different sediment types, and show up as repetitive, ~1 m thick layers deposited along the Spanish coastline about 6.7 million years ago. The sediments reveal a marine environment alternating between times when the ocean had intense biological productivity and thus high concentrations of fossil remains, and times when it did not, and the sediments were largely silts from nearby rivers. These alternations are thought to be linked to changes in temperature and wind circulation coinciding with the subtle but important variations in the precession cycle. Wind patterns alternatively churned up the marine basin and returned nutrients to the surface for more biological action and fossil production, and then slackened and allowed the dominance of river silt deposition. A remarkable thing about this is the precession signal becomes the geologic clock, and with very little in the way of age dating, scientists can count off the 1 m layers, 23,000 years at a time.

Figure 2.7 ***Layered Marine Sediments of the Sorbas Basin, Almeria, Spain, Deposited in the Proto-Mediterranean Ocean 6.7 million Years Ago.*** These layers in coastal marine sediments are manifested by color differences, corresponding to alternating intervals when the ocean had substantial marine productivity and thus abundant microfossil deposition, and intervals when it did not. Each layer, about 1 m in thickness, represents one precessional cycle of 23,000 years, and is driven by climate change–related variations in wind strength in the region. *Photo courtesy of J. Abel Flores.*

Thus, the main trigger of the glacial cycles that we have observed over the past 2.8 million years is variations in orbital parameters and thus the amount of energy the Earth receives. The cycles themselves are not symmetrical, and they exhibit far more variation in temperature than can be explained by solar energy alone. The aspects missing from the direct external driver of climate change are myriad intrinsic factors on Earth's surface, including variations in ice cover, ocean circulation, and atmospheric chemistry (https://www.carbonbrief.org/explainer-how-the-rise-and-fall-of-co2-levels-influenced-the-ice-ages). All of these factors are more immediate and localized than the methodical repetition of energy variations dictated by orbital variations, and indeed are all in play now and into the near future as AGW proceeds.

Intrinsic drivers of climate

Many aspects of climate and climate change have very much to do with what goes on in the atmosphere and at the surface of our planet.

This is seen most clearly in anthropogenic global warming. Whereas none of the external drivers act fast enough to impact our climate appreciably over the past 150 years, climate certainly has changed, and markedly so. This is due to several of the intrinsic drivers of energy balance (and thus climate) on Earth, including most importantly the substantial increase in heat-trapping gases emitted by human activity, as well as changes in Earth's reflectivity and its ocean circulation patterns. These then drive global and global changes in climate that we have seen since human activity became dominant with the Industrial Revolution of the late 1800s. All of these intrinsic drivers (and several more) have always existed and changed climate, but are simply changing much more rapidly than before industrial humans came on the scene.

Feedbacks, linearity, and albedo

One of the confusing factors with intrinsic drivers is that they are highly interactive, so it is sometimes difficult to sort out their relative impacts. Unlike an external driver like orbital variation, which is a one-directional climate driver, the intrinsic factors are affected by the external driver and by other internal factors via a variety of feedback loops. Briefly, feedback loops are connections between factors in a process that can either be positive, neutral, or negative. In a positive feedback, an increase in one factor might cause an increase in another factor, which then in turn causes an increase in the first factor, amplifying the original "push" to the system. And to make matters more complicated, some feedbacks are linear, and some are nonlinear. These feedback loops, and linearity—nonlinearity, are best understood by an example, for which we will use a glacier.

Consider a glacier, which is effectively a "river" of ice that flows from a source, where it is cold and has some snow accumulation, down to a terminus—think those tremendous pieces of glacier breaking off the cliff face at the seas edge in Alaska. A glacier is pretty hard to melt. Even when the potential temperature above a glacier should be above 0 degree, it is not in fact 0 degree. This is because of the most basic property of physics—color. For example, black materials appear black because the chemicals or minerals in that material absorb light over the full spectrum of colors, which are the optical manifestations of wavelengths. The fact that they absorb light, and hence the energy from that light, is well known intuitively to us. Walking across black asphalt in the summer can feel like walking across hot lava. In contrast, white materials reflect much of the light that they received, a characteristic well known to many sun-burned skiers every year as they are bombard by sunlight both from above their heads and bouncing off the very snow that they are enjoying.

Ice and snow are white—they reflect a lot of the light they receive. In fact, they reflect almost ¾ of the energy they receive from sunlight right back off the surface and back out to outer space, before anything has a chance to capture that potential solar energy. That is unless you happen to be snow skiing, in which case your skin can at least catch some of the ultraviolet rays bouncing off the snow. This provides the double whammy effect when you snow ski of getting sunburnt from above and from light reflecting off the slopes from below.

This balance between absorption and reflection has a scientific term, of course—albedo, which is a measure of how much light energy reflects off a surface. Something that absorbs 100% of light it receives, meaning that it reflects 0%, has an albedo of 0, whereas a material with the opposite characteristic has an albedo of 1. There are various mathematical reasons why the 0—1 scale is more useful than the 0%—100% scale.

Ice has a high albedo—about 0.7. So when sunlight hits ice, much of the energy is reflected right back out to outer space. When sunlight hits asphalt, which has a low albedo of about 0.3, much of the energy is absorbed, and the asphalt gives it off by heating the air above it. When sunlight hits ice, the air above the ice is actually cooler than would be expected compared to average absorption. In this way, ice inadvertently "protects itself" from melting, even when it really should be melting based on the amount of sunlight it is receiving. The air temperature has to increase quite a bit around ice to overcome this physical property of energy absorption and begin melting. The ice uses physics as a form of life insurance!

These are all relatively straightforward applications of basic principles of physics, but with a very intriguing implication for climate change. As air temperatures are warming because of human activities, they have begun exceeding the ice's capacity for self-protection, and ice worldwide has begun melting. Melting of an ice sheet is actually a very nonlinear response. A linear response is a response proportional to a force—think of pushing a glass across your kitchen table. A gentle push will move the glass a short distance, and a harder push will move it farther. The distance that it moves is linearly related to the energy exerted in pushing it. The nonlinear response comes when the glass reaches the edge of the table. At this point, even a gentle nudge will send the glass over the edge, resulting in it moving a great distance, at a great speed, as it crashes to the floor. The slow warming causes some melting at the edges of a sheet of ice, the linear response, but as the edge of this ice sheet begins to erode, it loses its high albedo (assume that there is rock underneath, with a low albedo so high heat absorption capacity). This causes the air on the edge of the eroding ice sheet to heat up even more than it would otherwise, causing increased rates of melting, more exposure of rock, more absorption of heat, more melting, and so forth.

Earth systems do tend to be residing in states of stability. The cup example typifies these stable states. One state is on the table, and the next is on the floor. Both are stable horizontal planes, in this case, but the interval between the two is not. The same occurs with glacial cycles, in that it is difficult to get them out of either their icy glacial state or their ice-free interglacial state. Each will resist a push until they reach a tipping point, at which the transition occurs quickly. In the context of AGW, global ice cover is a glass that we have inadvertently moved close to the edge of the global table. Continual gradual increases in air temperature are the gentle push that is causing ice sheets to tumble off the table, crumbling and melting and retreating the world over. We do not know exactly there the table edge is—in other words, how much more heating ice can take before "going nonlinear", but based on recent satellite observations, we think that we are almost there.

Greenhouse gasses

The Earth's current atmosphere is made up of about 78% nitrogen, 21% oxygen, and 1% of a host of other gasses. As will be discussed in great deal in other chapters, the past atmosphere was quite different, not just with the nitrogen and oxygen components, but also the gasses that comprise the other 1%. Several of the gasses that are part of this 1% are actually extremely powerful drivers of modern, and past, climate. Carbon dioxide, water vapor, and methane all have a unique quality and serve as heat-trapping gasses due to their unique physical properties (https://climate.nasa.gov/faq/19/what-is-the-greenhouse-effect/). Neither nitrogen nor oxygen has these same characteristics. Thus, although past concentrations of oxygen, and to a lesser extent nitrogen, have varied significantly through geologic time, they had little direct influence on past climate. On the other hand, variations in carbon dioxide and methane have had a huge influence on climate, past, present and future (https://www.epa.gov/climate-indicators/greenhouse-gases). Water vapor is not included on that list, for reasons that will become clear shortly.

Much of this book is about the interplay between greenhouse gasses like carbon dioxide, earth processes, and life through time. But to get an understanding of why these minor gasses are such powerful drivers of climate, it is critical to understand their physics and that of sunlight. The Sun emits radiation across a broad spectral range. Much of this radiation is in the visible (to our eyes) range of wavelengths, from red to violet.

The Sun also emits radiation at shorter wavelengths than our eyes can see, in the ultraviolet range. Interestingly, bees can see in this range, and in fact pursue certain types of flowers for nectar because they also reflect light in that range. It is common knowledge that ultraviolet radiation is very powerful and very harmful to biological cells, like those in our skin—hence the need for sunscreen. This absorption occurs because of gasses like ozone and water vapor, as well as aerosol particles. The Sun also emits radiation in the longer infrared wavelengths, which do not have the dangerous impacts of short wavelength ultraviolet radiation and are felt most directly as heat energy.

Of the total amount of radiation that arrives at the top of our atmosphere, about 25% is reflected directly back out to outer space by clouds or aerosol particles, another 20% is absorbed by the atmosphere, and the rest hits the Earth's surface. Radiation that hits the surface of the Earth can either be absorbed or reflected (Fig. 2.8). The balance of absorbed to reflected sunlight relates to the albedo effect discussed earlier. The reflected radiation is always at a longer wavelength than incoming radiation, and thus much of it is infrared heat. Thus, although absorbed sunlight heats the Earth's surface directly, reflected sunlight plus that absorbed by the atmosphere directly both contribute to the heating of the atmosphere, which in turn can warm the surface. The gasses most responsible for trapping that atmospheric heat are water vapor, carbon dioxide, methane and ozone, and they do so because they are particularly reactive to energy in the infrared wavelengths.

What is perhaps less known is that our atmosphere does a reasonable job of directly absorbing much of the ultraviolet radiation emitted from the Sun, particularly the extremely high energy and biologically dangerous UVC and UVB bands. Ozone is a gas comprised of three molecules of oxygen bonded together. In the lower atmosphere, ozone is produced by interactions between sunlight and volatile organic compounds and is quite harmful to human lungs. In the upper atmosphere, however, ozone is produced by interactions between regular oxygen gas (two oxygen atoms bonded together) and ultraviolet radiation, which is effectively absorbed in the process.

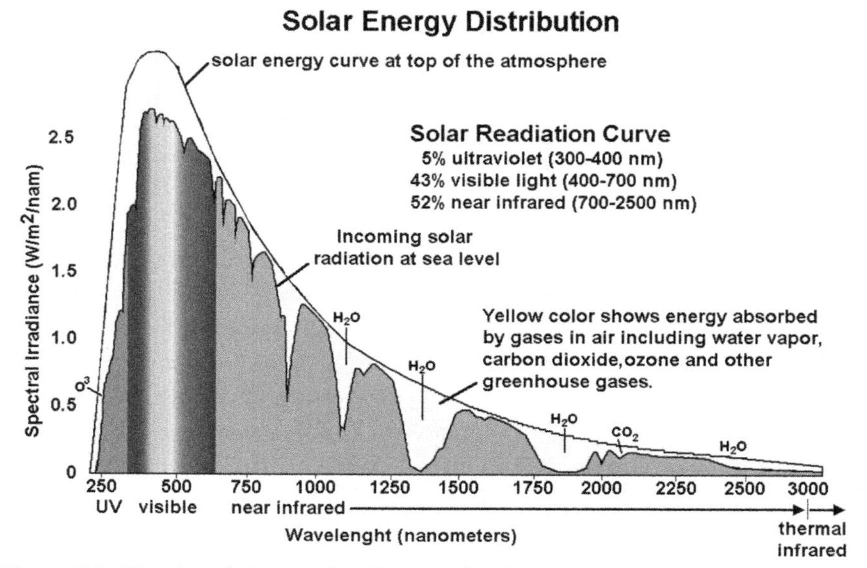

Figure 2.8 ***Wavelength Energy Distribution of Solar Radiation.*** Solar radiation enters Earth's atmosphere in a skewed bell curve pattern including ultraviolet, visible, and infrared wavelengths. This energy spectrum is modified by gases and aerosols in our atmosphere. The portion of incoming radiation that is retained by the Earth, either through direct heating of the surface or through atmospheric absorption, is critical to sustaining life on the planet. *From https://geologycafe.com/oceans/images/insolation_curve.jpg.*

As is covered later in this book, the formation of the thick layer of upper atmosphere ozone is relatively "recent" in Earth's long history, occurring about 600—700 million years ago, but it is a critical first shield to protect organisms from the damaging, mutagenic effects of ultraviolet radiation should it make its way to the planet's surface. This is why scientists were so alarmed to see this protective barrier thinning out in the 1970s (the first measurements were taken then, but certainly it had begun thinning before then), and rushed to find the cause. Atmospheric chemists Mario Molina, F. Sherwood Rowland, and Paul Crutzen pinpointed the human-produced molecule chlorofluorocarbon (CFC) as the culprit, and won a Nobel Prize in Chemistry in 1995 for doing so. CFCs are gasses that were released from a wide variety of products, most notably Freon coolant, but also cleaning agents and foam manufacturing. They did not appear toxic in initial studies, and thus were approved for use, but their toxic effects to the protective ozone layer were not identified until Molina, Rowland, and Crutzen put together the pieces, and ultimately helped to lead the signing of the Montreal Protocol, a successful international ban on CFC production in 1987. Since CFCs have a lifetime of about 200 years in the upper atmosphere, it will take some time for the ban to end up "curing the ozone hole," but meanwhile, CFCs have another important impact—they are also among the most powerful greenhouse gases that exist owing to their capacity to absorb a tremendous proportion of radiation. It is sometimes wrongly stated that the ozone hole causes global warming—it does not, but it is indicative of the presence of CFCs, which certainly do contribute to atmospheric warming.

Returning to greenhouse gasses in general, they effectively control the Earth's radiative balance—the net balance between energy received at the top of the atmosphere and energy returned to outer space. The physics of these gasses is straightforward, in that they individually have the capacity to absorb a certain portion of radiation that passes "through" them. Their role in the atmosphere has been unchanged since the Earth formed 4.6 billion years ago—the higher the amount of greenhouse gasses in the atmosphere, the greater proportion of solar energy that is retained in the atmosphere (and vice versa). This effect can be measured directly, as we have been doing for the past 60+ years but have hypothesized to be the case since the 1800s. It is thus a puzzle that so many deny the basic laws of physics, which work on Earth and every other planetary body with an atmosphere, and claim that the observed increase in greenhouse gasses will not cause AGW. This general topic—the future of climate change and the life on this planet—will be detailed in the final chapter, but suffice it to say that greenhouse gasses do trap heat on the planet, and always have, and always will.

The different greenhouse gasses have different heat-trapping efficiencies based on their molecular physics and play different net roles in trapping heat. The efficiency of the molecule to trap heat, or its global warming potential, is typically measured in comparison to that of carbon dioxide over a certain length of time. For example, the gas methane is about 28 times more effective at trapping heat than is a molecule of carbon dioxide over 100 years. In other words, it would take 28 molecules of carbon dioxide to trap as much heat as only one molecule of methane over a century. Interestingly, the methane molecule itself is much more efficient at absorbing heat than that in comparison to carbon dioxide, but only lasts about 10 years in the atmosphere, compared to carbon dioxide's 100 years. On top of that, methane is removed from the atmosphere by oxidation, but this process simply converts the methane to carbon dioxide, where it continues to trap heat. Given methane's global warming potential, it would seem that this is the dominant greenhouse gas, but this potential is tempered by how little methane there is—it is about 200 times less abundant than carbon dioxide, and thus, although much more efficient as a heat-trapping gas, it contributes only about 17% to the total atmospheric radiative effect. CFCs have a global warming potential of up to 10,000 times that of carbon dioxide over a century, but because they are present in only small amounts, they contribute far less than 1% to Earth's radiative forcing.

The role of water vapor as a greenhouse gas and its global warming potential are actually tricky to calculate, but in practice it does not matter in a broad sense. It certainly absorbs infrared radiation to a greater extent and over a broader range of wavelengths than carbon dioxide, but it is quite variable across the globe, is driven by major downward fluxes via precipitation and upward fluxes via evaporative processes, and does not decay or breakdown in the atmosphere. But this particular greenhouse gas is strictly controlled by temperature—a warmer planet drives more evaporation versus precipitation, leading to a higher humidity in the atmosphere and thus a higher temperature (and vice versa). There is clear evidence for this process in action in the geologic record, where ice cores reveal that cold glacial climates lead to expansion of desert zones and greater amounts of dust in the atmosphere. In this case, the cores contain chemical markers for cold temperatures and mineral markers in the dusty layers that coincide with the cold temperatures. So rather than being a direct greenhouse gas, it can be considered an indirect one, amplifying the effects (via positive feedbacks) of external and other intrinsic drivers of climate change but not controlling the trajectory of change.

Sources and sinks for greenhouse gases

Because the concentration of climate-controlling greenhouse gases in the atmosphere changes over time, they clearly have sources that produce them and sinks that consume them. These become part of the cycle of the particular greenhouse gas and are typically in some sort of varying balance through time. Take the cycle for present-day carbon dioxide, for example. The iconic record of atmospheric carbon dioxide in air from Mauna Loa, Hawaii (Fig. 2.9), reveals key aspects of this balance—one that is a clear reflection of human activities and one that is part of an ecosystem process that has functioned for millennia.

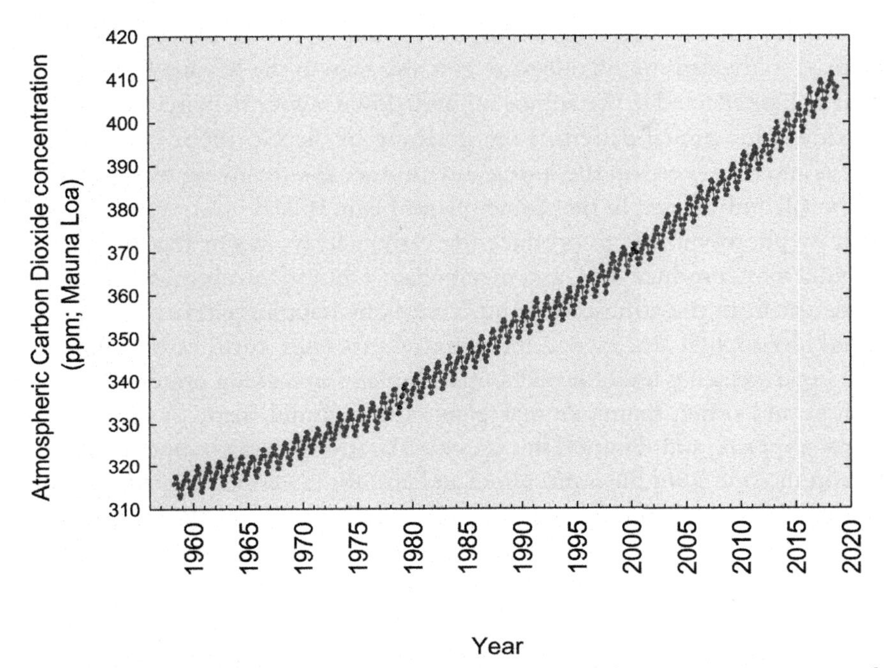

Figure 2.9 ***Concentration of Carbon dioxide in the Atmosphere.*** The concentration of carbon dioxide in the atmosphere has been measured continuously from 1958 at a research station atop Mauna Loa, Hawaii. https://earthobservatory.nasa.gov/images/43182/mauna-loa-observatory. This record provides evidence for the human influence on the global carbon dioxide cycle (long upward trend) as well as the role of seasonal changes in northern hemisphere vegetation plays in carbon exchange (annual wiggles). *Data from NOAA. https://www.esrl.noaa.gov/gmd/ccgg/trends/.*

Human activities have accelerated the sources of carbon dioxide into the atmosphere at such a great rate that ecosystem and geologic factors that would provide the offsetting sinks are overwhelmed. This results in the general increase in the concentration of carbon dioxide in the atmosphere that is observed since the beginning of this record in 1958, although certainly this increase began earlier, with the intensification in the use of coal, gas, and oil at the start of the Industrial Revolution. These sources have largely tapped vast amounts of geologically deposited carbon that accumulated in organic matter over millions of years, and converted them rapidly over only a century into the by-product of burning organic matter—carbon dioxide. Other human-related sources of carbon dioxide exist and will be discussed in detail in the last chapter, and collectively, they have resulted in a 30% increase in the global concentration of this gas in the atmosphere, and thus largely drive the AGW that we observe today.

One of the key sinks that is currently being overwhelmed by the rapid increase in the carbon dioxide sources is also seen in the Mauna Loa record. This sink is reflected in the annual up-and-down sawtooth pattern of carbon dioxide. This annual pattern is the greening of the Northern Hemisphere forests and ecosystems in the spring and summer and the dying off and decay in the fall and winter. In the spring, plants begin to leaf out to capture sunlight by photosynthesis to produce the carbon-heavy sugars that they need to build mass, produce fruit, and metabolize. This process absorbs carbon dioxide gas from the atmosphere and, via a light-trapping pigment molecule called chlorophyll and myriad biochemical processes, turns carbon dioxide into sugar, which is used for metabolism but also in making proteins, waxes, lignins, and other chemicals that plants use to build mass, or reproduce. Through spring and summer, this activity is so strong that it temporarily pulls carbon dioxide from the atmosphere and absorbs in ecosystems. As summer turns to fall, the shortening daylight triggers plants to begin closing up shop, as it were, including stopping photosynthesis, dropping leaves, and, for annual plants, dying after producing seeds. The dead plant matter drops to the soil, where the vast bacterial and fungal ecosystem there hungrily consume the residual sugars and carbon-dense organic matter and release this carbon back to the atmosphere as carbon dioxide. This process is repeated year after year, decade after decade in the seasonal pattern that has existed since life developed on this planet almost 4 billion years ago.

As the Mauna Loa record shows, this source–sink pulse of the biosphere accounts for variations of only a percent or so in the atmospheric carbon dioxide. And curiously, this sawtooth pattern is only present in the Northern Hemisphere records. One would expect to see this in records from the Southern Hemisphere as well, but with the "downs" representing their opposite seasons. Instead, there is only the steady increase from human activities, with only very minor annual variations. This is due to the interesting current configurations of our continents, which is very "Northern Hemisphere heavy." Recall from Fig. 2.6 that the Arctic Circle passes over almost all continent on a planet that is 70% ocean—the inverse is true of the Antarctic Circle, which is nearly completely over ocean. Indeed, the Southern Hemisphere as a whole has very little land mass compared to the Northern Hemisphere, and much of this landmass—almost the entire continent of Australia, for example—resides at a latitude that is relatively arid. Thus, very few forests exist in the Southern Hemisphere, and the atmospheric carbon dioxide records are pretty stable on an annual basis. Carbon dioxide does mix between the hemispheres, but on the time scale of atmospheric mixing, which is longer than a year.

Many other nonhuman sources exist for carbon dioxide, and we will cover most of them in the course of this book. Briefly, these sources are typically of a much less dynamic nature than the annual pulse of greening and decay, or the rapid injection of carbon dioxide from human activities. One source is the very slow leaking of carbon dioxide from volcanoes that are present on land and particularly under the ocean—this carbon dioxide is related to Earth's dynamic interior dynamo, which cycles carbon down into the deep Earth and back up again on times scales of millions of years. This source certainly varies over time, but is only influential for longer-term changes in atmospheric carbon dioxide and climate change. Another source is the bubbling up of carbon dioxide from the ocean as the ocean continues its slow global circulation pattern. The carbon dioxide is produced by bacterial decay of organic matter in the ocean, much in the same way as the seasonal decay of plant leaves and stems occurs on land. The difference here is that the carbon dioxide is released into deep water and is held there until the water eventually surfaces. This pathway of ocean circulation is also slow compared to the seasonal and human-influenced processes, but is much faster than the deep Earth carbon cycle. On average, a given molecule of water that is on the surface of the ocean will be entrained into a deep ocean flow pattern and eventually emerge about 1000 years later. During that entire time, the water does not interact with the atmosphere, so any changes in the body of water will accumulate over time until it upwells and releases its dissolved carbon dioxide into the atmosphere.

The processes that are the sinks for carbon dioxide out of the atmosphere are simply the inverse of the sources discussed above. There is the annual uptake on land during the growing season, there is the uptake of carbon dioxide by photosynthetic organisms in the ocean and the settling of that material to the depths (where it is slowly consumed by bacteria), and there is the general removal of carbon from the Earth's surface system through slow geologic processes. One caveat about this latter process is that whereas carbon dioxide is released from the Earth's interior via volcanic emissions, the absorption and eventual removal of it into the deep Earth cycle is a bit more complicated. Carbon dioxide is absorbed from the atmosphere via the weathering of minerals on land, is then transported by rivers in dissolved form, as carbonate ions, to the oceans. That carbonate is taken up by microorganisms that make carbonate shells, those shells eventually settle to the seafloor as limestone mud, and this material is transported to the Earth's interior through a geologic process called subduction. Regardless of these complications, the millions of years timescale of this cycle are the same.

Based on these various aspects of the carbon cycle, we consider it to have a short-term cycle, a medium-term cycle, and a long-term cycle. The short-term cycle functions on the timescales of tens to hundreds of years and involves exchanges between the atmosphere and various components of the Earth's ecosystems, ranging from the seasonal greening phenomenon to slower processes related to forest growth, soil exchange, and other biotic reactions. Sources, sinks, and feedbacks in this short-term cycle function on the timescale of generations, and thus these are of strong interest to researchers in understanding how human activities have perturbed these systems, and what to expect in the near future from these interactions. The medium-term cycle for carbon is tied to ocean circulation patterns and ocean chemistry and operates at the scales of thousands of years. So, not within a lifetime of an individual, but certainly within the span of recorded human history. The long-term geologic cycles are those of profound interest to how the Earth sets long-term trajectories in climate, on the order of millions of years.

Methane is of course also an important greenhouse gas with its own unique sources, cycling, and sinks. Although secondary in terms of net climate contribution to carbon dioxide, there have been times in the geologic past when large perturbations in the methane cycle did seem to be in the climate drivers' seat, causing massive changes in global ecosystems and geochemical cycles. One such perturbation, which occurred about 55 million years ago and marked the boundary between two geologic periods, is called the Paleocene–Eocene Thermal Maximum (PETM). Details on the chronology, impacts, and recovery of the PETM will be thoroughly described later in the book, but in a nutshell, so much methane was released from a yet-to-be-confirmed source that the planet warmed several degrees, and the oceans acidified and many ecosystems collapsed as methane was oxidized to carbon dioxide, and that carbon dioxide entered the ocean and raised the acidity level.

Methane is produced by a number of processes, all with the same characteristic—the environment of production has very little, or no, oxygen. One such environment is in stagnant ponds and saturated soils, such as found in wetlands and rice paddies, where all of the oxygen normally present is used by bacteria that decompose organic carbon. This anoxic setting then promotes a number of other bacterial groups that do not need oxygen to survive, including one that generates methane (a so-called methanogen). Conversely, there are other bacteria that consume this methane waste product, called a methanotroph. Another environment where methane is produced is in some ocean sediments, where the released methane is not consumed but instead is trapped in a methane-water ice called a methane clathrate which is stable in deeper, colder environments but rapidly melts when brought up to the surface, releasing its methane (Fig. 2.10). It may also melt in the sediments themselves if the overlying water warms, a significant concern for enhancing AGW as a positive feedback (warming air = warming water = methane release from clathrates = more warming = greater area of warm water = more methane released...). Methane is also produced in the intestinal tracts of ruminant animals, the group of split-hooved animals, such as giraffes, llamas, and zebras, but also many

Figure 2.10 ***Methane Clathrate.*** Methane clathrate is a methane-ice mineral that forms in ocean sediments. It is very sensitive to changes in pressure and temperature, and melts when brought to the surface. It burns at such a low temperature, in fact, that the flaming clathrate can be held comfortably in the bare hand. Clathrates are considered one of the wild cards in anthropogenic global warming, as when they begin melting they rapidly destabilize and could release a tremendous amount of warming methane worldwide.

domesticated livestock like cows and goats. The domestication and consumption of ruminants has significantly increased the human-related sources of methane to the atmosphere, as has the cultivation of rice as a main cereal grain for large parts of the world. Methane sinks include methanotrophs, but the most important one is simply atmospheric oxygen, which converts methane into carbon dioxide on timescales of years.

Clouds

Clouds are masses of mainly water droplets suspended at altitude at a quasi-stable temperature and pressure. When these conditions change, water droplets can get smaller, turn into water vapor, and result in the cloud disappearing, or combine into larger droplets, exceeding the stable conditions for that elevation and drop from the cloud as rain, snow, or hail. And of course, clouds come in myriad shapes and sizes and densities, and can be present at various altitudes. With all of the diversity in cloud shape, density, height, and longevity, it is probably not surprising that their impacts on climate are equally complex. Indeed, "the cloud problem" is one of the greater challenges at building climate models that will forecast into the future. Simply put, clouds can act to reflect incoming solar radiation back out to space causing cooling, trap reflected solar energy in the atmosphere causing heating, or both. Indeed, the role of clouds even varies during a given day—low level cloud cover, which is prevalent over the oceans, tends to reflect sunlight back to space during the day, while trapping heat during the night. One of the biggest current debates about the role of clouds is on future climate change. How dense will clouds be in a warmer world? How then will they provide positive, or negative, feedbacks to increased warming? A study in 2019 (https://e360.yale.edu/features/why-clouds-are-the-key-to-new-troubling-projections-on-warming) attempted to answer that question, and the resulting model indicated that cloud density will be greatly reduced (Fig. 2.11), thus adding a positive feedback to warming (https://e360.yale.edu/features/why-clouds-are-the-key-to-new-troubling-projections-on-warming). Suffice it to say, unlike the clear physics of most of the external and intrinsic drivers of climate change, the role of clouds remains a bit of a scientific mystery.

Aerosols

Unlike clouds, aerosols (fine particulates in the atmosphere) largely act to cool the planet by reflecting incoming solar radiation back to space

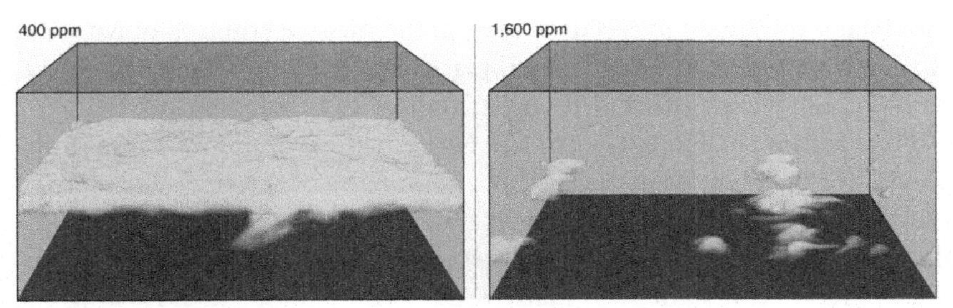

Figure 2.11 ***Model Visualization of Cloud Cover and Climate Change With Current Carbon dioxide (Left) and High Carbon dioxide Associated With Extreme Warming (Right).*** - One challenge to understanding the role that clouds play in climate and future climate change is that they both reflect sunlight, causing cooling, and trap heat, causing warming. The balance between the cooling and warming effects is strongly related to cloud thickness and height, which vary over time. Climate modelers had long thought that increased global temperatures with AGW would result in more evaporation and more cloud cover—recent results argue the opposite, and indicate that the thinner clouds will allow much more solar radiation to hit the Earth under higher carbon dioxide concentrations, thus creating a positive feedback loop. *From https://doi.org/10.1038/s41561-019-0310-1.*

(https://www.nature.com/scitable/knowledge/library/aerosols-and-their-relation-to-global-climate-102215345/). Although a few aerosol types exist that do have heat-trapping abilities, like soot, most act to cool the planet (one caveat is that aerosols can contribute to cloud formation as "cloud condensation nuclei." In this role, the net climate impact of aerosols again depends on the location, density, and height of clouds.) Indeed, this cooling effect can be profound, and is clearly present in the geologic recorded and recent history. The massive asteroid impact that killed off the dinosaurs about 66 million years ago, to be covered in some detail in a later chapter, ejected a tremendous amount of debris and dust into the atmosphere, including the stratosphere, where particles remain for several years before eventually settling out. These aerosols cooled the planet tremendously and resulted in a "nuclear winter" scenario that dimmed sunlight and cooled the surface to such an extent that ecosystems collapsed, and many of the hapless survivors of the impact itself subsequently died of starvation. Much more recently, the massive eruption of Mount Pinatubo in the Philippines in September 1991 ejected 15 million tons of sulfur dioxide into the stratosphere, which caused secondary aerosols to form. Over the 15 months posteruption, these reflective aerosols dropped the average global temperature of about 1°F (0.6°C) until they eventually rained out (https://earthobservatory.nasa.gov/images/1510/global-effects-of-mount-pinatubo). This particular eruption is not nearly the largest in recent history, as the massive eruption of Mt. Krakatoa in 1883 also reduced global temperatures significantly, and caused tsunamis and ash clouds that killed tens of thousands of people.

The climate change impact of the Mt. Pinatubo eruption is one of the catalysts for an effort to recreate this phenomenon to reduce the impacts of carbon dioxide increases and AGW, by intentionally "seeding" the stratosphere with sulfur dioxide. This process, one of several "geoengineering" fixes to AGW, is fraught with many issues (who pays for it? how long can it be sustained? what are in the unexpected consequences of injecting chemicals into the stratosphere on this large of a scale), but reveals how important aerosols can be in the short-term climate system. One of the ironies of the development of renewable and noncarbon energy sources is that AGW is caused in large part by the combustion of fossil fuels, a by-product of which are various gasses that form secondary aerosols over cities and industrial regions. These reflective clouds of smog actually have a cooling effect, masking some of the expected warming. Efforts to improve the environment and human health in general by removing fossil fuel consumption from the energy and transportation mix will yield cleaner air, and thus higher temperatures over industrial and populated regions

How do we know about climates of the past?

Many different tools have been developed and deployed to study past climate change, and more are being formulated and proposed every day. They typically are only effective for certain time intervals, or different climate variables. None of these techniques is perfect, and climate scientists tend to be very careful about how they interpret their data to take into account these uncertainties. Furthermore, some techniques are designed to specifically measure a climate variable like temperature, or rainfall. Others are designed to examine indirect aspects of climate change, such as climate impacts on chemistry, or ecology, or erosion—many of these are called proxy records, as they record one variable that indicates changes in another. It is helpful to have an understanding of the scope and limitations of these records of past climate change, especially given how much the science evolves, and how quickly. This section arranges the techniques and records from recent past to distant past and focuses less on the techniques of age dating and instead examines the proxy records themselves.

0–5000 years ago

Historical thermometer records of a global nature extend back to the late 1800s, with less representative (and thus less accurate) values in the early part of the record and much more accurate values toward the present. Although individual thermometers were invented and used sporadically before this time, we really only have valid global land and ocean temperature records for the past ~150 years. These values have to be adjusted by local factors near the temperature monitors—for example, if a thermometer station was originally installed in an open farm field which later was encroached on by development, the temperature would go up due to the greater absorption of solar energy on built materials (i.e., local buildings, asphalt drives, etc.). This apparent temperature increase needs to be adjusted downward when generating temperature trend records to account for these changes. Given the global coverage that satellites provide, they have been indispensable at augmenting ocean temperature moorings, and "filling in the gaps" over vast areas of the ocean. The satellites are indirectly measuring temperature by infrared radiation emissions, but are ground-truthed using the monitors at the ocean surface.

Similar to temperature, we have robust direct precipitation records since about the late 1800s, although again it is much more concentrated on land via rain gauges than over the ocean. More recently, extensive satellite coverage coupled with land-based radar measurements has substantially increased the spatial scale of the precipitation data. As with temperature, ocean records of precipitation are more limited in spatial resolution, but satellite coverage significantly increased measurements over the ocean.

But several indirect records of both temperature and precipitation are available in the recent period, with various limitations in their spatial coverage and reliability. One such documentation is simply the historical written record. There is some form of written record over the past 5000 years of human history, and not unsurprisingly, one topic that people have always written about is the weather! These are individual accounts at specific locations, and tend to be biased toward the extreme event—floods, freezing weather, storms, etc. Starting with the "birth" of the field of meteorology, represented in Sanskrit by the Upanishads in India 5000 years ago and continuing with writing from Greece, Rome, and throughout the world since the end of the Middle Ages, these documents reflect, with varying accuracy, various weather events. Drawings and paintings do as well, with the Little Ice Age, a brief interval of cold climate immediately preceding the Industrial Revolution, displayed in paintings of glaciers in Europe covering areas where they were not subsequently, and George Washington crossing an abnormally icy Delaware River, as two examples.

But perhaps more "scientific" than the written record over the past 5,000 years are various climate archives stored in tree rings and in lake sediments. The amazing Bristlecone Pine (*Pinus longaeva*) is the oldest known organism on the planet, with some individuals coming in at almost 5000 years old (https://www.newyorker.com/magazine/2020/01/20/the-past-and-the-future-of-the-earths-oldest-trees). This longevity alone is impressive, but for climate scientists, it is also a great boon—because trees record annual growth rings, and the width of the rings is a function of climate (rainfall and temperature); these amazing sentinels serve as an annual measure of climate. These obviously have limitations, in that they grow in only a few regions and the thickness of the growth rings is a relative measure of climate only, but they also record global events if severe enough. For example, so-called frost cracks, when the annual ring growth was impeded by extremely cold weather, reveal some extreme events that can be correlated globally. Most of these seem to be massive volcanic events, when aerosols emitted by eruption cooled the planet for 1–2 years. Lake sediments also can have annual layers, called varves, which can literally be counted backward from the present.

The composition and thickness of these layers can be influenced by particular storm events that wash a significant amount of sediment into the lake. Additionally, the geochemistry and biology of lakes can indicate temperature of the lake surface water, which reflects air temperature (https://www.upi.com/Science_News/2019/09/04/Sediment-from-Europes-oldest-lake-reveals-1-million-years-of-climate-history/3261567542417/http://www.iodp.org/). This is recorded in the species of organisms present over time, and/or the chemical composition of various components within the sediment. For this latter, one sensitive tool that is used in many settings is the isotopic composition of materials, including fossils. For climate, the most critical one is the value of oxygen isotopes trapped in shells. Oxygen has three isotopes, or chemically similar versions, of itself, and the ratio of the heaviest to the lightest is strongly influenced by temperature. Thus, a record of heavy-to-light isotopes from fossils through time can give an indication of relative changes in water temperature and thus climate. More sophisticated tools exist as well, including examining the chemistry of the organic matter itself, and these tools are increasingly applied to reconstructing past climates.

5000–1,000,000 years ago

In the interval stretching back from the beginning of written history 5000 years ago, through the evolution of our species 200,000 years ago, and back one million years ago, a new set of tools and proxy records exist, including archives in ice sheets, lakes, and the ocean. The hallmark for documenting the past concentration of carbon dioxide in the atmosphere, and past temperature, is certainly cores taken from ice sheets. Ice sheets at high latitudes accumulate a new layer of ice every year. This ice starts actually as snowfall, and as the snow solidifies, it locks in bubbles of the atmosphere. The atmosphere in these bubbles represents the atmosphere at the time of trapping, and remains stable for hundreds of thousands of years. Over the past several decades, a series of expeditions have ventured to areas with substantial continental ice sheets, such as Greenland and Antarctica, and have drilled boreholes into the ice sheets and extracted cores of ice (https://icecores.org/about-ice-cores), going back as far as 800,000 years ago in Antarctica (Fig. 2.12). These cores are carefully age dated by the annual counting layers or other radioisotopic measures on the ice or particles of dust or ash that are present. But the key motivator of the ice coring itself is that the core can be cut into thin layers, placed in a special chamber, and melted, thus releasing the ancient atmosphere trapped in the ice bubbles

Figure 2.12 ***Ice Core.*** This ice core, from the West Antarctic Ice Sheet Divide, shows a dark layer of volcanic ash (which can be used for accurate dating) and annual layers of slight more opaque and less opaque ice. This facility is the U.S. National Science Foundation Ice Core Facility (https://icecores.org/about-ice-cores).

for analysis of its gas content. In essence, the instruments are "breathing" air that might be as old as 800,000 years. This process gives a direct measurement of, say, the concentration of carbon dioxide in that ancient atmosphere. The isotopic composition of the melted ice itself also gives an indirect measure of the temperature at the time of formation. This information has yielded valuable insights into the orbitally driven glacial cycles over the past million years and has shown that the concentration of greenhouse gasses provides strong additional forcing of global temperatures above and beyond that predicted by orbital variations alone. The results have also placed our current AGW in a stark perspective—during all of the previous interglacial intervals of the last million years, the maximum carbon dioxide concentration was about 300 parts per million, and we are now at 418 ppm, and climbing.

Additional information can also be extracted from ice cores, particularly the dust that is trapped along with the air. Some of this is volcanic ash from eruptions, and this proves invaluable for providing absolute age markers in the ice core—the chemistry of volcanic eruptions is typically well known, and thus one particular ash can usually be traced back to the age of a particular eruption event. Additionally, a higher concentration of atmospheric dust found in ice from glacial periods points to a higher area of arid lands (deserts) and thus lower net atmospheric moisture in the atmosphere, as noted earlier. Even the presence of salts, blown in from nearby ocean areas, can give clues as to the salinity of oceans in the past.

Collectively, ice core techniques have opened up the recent glacial past in a way that no previous archive was able to do, but the record is very spatially restricted. Ice core records from the poles, and from some high alpine glaciers, certainly provide information about global greenhouse gas concentrations, but the temperature records are localized to those cold regions, and do not shed much light on temperatures in the tropics or the temperate zones. Additionally, they are all land-based, so they provide very little information about the impacts of climate change on lake or ocean temperature, circulation or ecology. For these purposes, cores of sediments from lakes or the ocean are much more informative, and give a more global coverage of climate impacts. Lake core records were discussed previously and are included here as lake sediment records can also extend to a million years ago. But marine sediment core records have opened up our understanding of ocean systems in the past and constitute a hallmark achievement in earth sciences.

The earliest systematic research of the ocean and seafloor sediments was conducted as part of the HMS Challenger expeditions, from 1871 to 1876. But drilling cores and extracting records of past marine environments began in earnest with the Deep Sea Drilling Project in 1968, a U.S.-based initiative which has nevertheless engaged with scientists from around the world, and transitioned into the Integrated Ocean Drilling Program (Fig. 2.13), with many global partners (it was renamed the International Ocean Discovery Program in 2013). The cores of sediment extracted from the seafloor all around the world have revealed the stunning impacts of orbitally driven climate change over the last million years. These records include variations in the isotopic values of shells from plankton, which, as in lake sediments, reflect the temperature of the seawater they formed in as well as document the amount of water locked up in ice sheets around the world. It is this latter observation that provides critical information about sea level changes in the past related to climate change—when ice sheets grow, they do so from precipitation. As most of the water in our atmosphere is originally derived from evaporation from the ocean, each layer of accumulating ice in an ice sheet is effectively removing water from the oceans. And it is not a trivial amount at all—at the peak of the last glacial event about 20,000 years ago, global sea levels were about 120 m lower than today. Indeed, marine archaeologists can find early human habitations at what was previously beachfront property, but is now submerged below 120 m of ocean water.

Figure 2.13 ***A Scientific Ocean Drilling Vessel.*** The JOIDES Resolution is a U.S.-operated ocean drilling vessel that has the capability of drilling and retrieving sediment cores for paleoclimate research from deep in the seafloor. This vessel is a sailing laboratory, with 20–30 scientists from all around the world onboard at any one time, and state-of-the-art research labs.

More than 1 million years ago

For time periods great than 1 million years ago, paleoclimatologists can still use ocean sediment records, but increasingly need to rely on the more sparse records stored in rock deposits and fossil beds on land. The oldest ocean sediment that exists is in the western Pacific Ocean and is about 170 million years old. With a total Earth age of 4.6 billion years, the seafloor is relatively young at 170 million years, and is continuously rejuvenated by the process of plate tectonics and subduction introduced in the first chapter. But not all of that ocean sediment gets recycled in the deep Earth—some is scraped off or otherwise trapped on the continents themselves. This occurs through plate collisions and mountain-building, for example, with the Himalayas a classic set of marine limestone and shales that were deposited on the seafloor 50–100 million years ago but are now at 6,000 m above sea level. Practically, the entire west coast of the United States is comprised of a series of volcanic arcs and marine sediment deposits that resisted subduction and were instead accreted onto the continent. In both cases—the Himalayas and the western United States—valuable fossil remains and other clues in the chemistry of the rocks point to significant changes in past climate and marine ecology.

The fossil remains of marine organisms, like shells and ancient coral reefs, can be exploited as climate indicators, by their species affinities (cold-loving, hot-loving, etc.), their chemistry, or both. But other types of fossils can also be present in these deposits—namely, those terrestrial organisms that were washed into ancient lakes or oceans via rivers. Land plants have their own climate story to tell, and as with fossils, their isotopic composition and growth habit can indicate the temperature, rainfall, or both at the site where they grew. Additional information can even be gleaned from tree leaves, for example—the density and size of the pores (called stoma) that perform gas exchange in leaves related to photosynthesis reveals to some extent the relative humidity and carbon dioxide composition of the atmosphere.

Additional information about climate in the distant past is reflected in the rock record. Glacial deposits can reflect the extent and timing of ice sheet movements on land. Sand dune deposits and loess deposits, which are layered dust deposits, can record the extent of deserts and thus indicates the relative rainfall patterns on land. Ancient lake deposits can record the climatic conditions in the watersheds when and where they were formed. As with all of these reconstructions, their accuracy at reflecting "true" climate conditions tends to get poorer the farther back in time they represent, and thus the fidelity of the particular determinant to truly define the climate parameter in question can be weakened.

This fidelity issue is best understood by comparing two of the proxy records formerly discussed. Air bubbles in ice cores record exactly the gas concentrations in ancient atmospheres, but the "lock-in" timing of the bubble being trapped in ice spans several years. In contrast, leaf stoma density and size reflect moisture and carbon dioxide in ancient atmospheres, but does so relatively poorly, and the age dating of the leaves themselves is sometimes quite poor. The Earth contains only imperfect records of its past, and thus challenges our understanding of past climate change and its relationship to life.

Summary

The physics of atmospheric motion and resulting weather is devilishly complex and chaotic, which is why weather predictions start falling apart about a week so into the future. Fortunately, climate and the various drivers of climate are more sensible, abiding by a series of rules dictated by the greenhouse gas composition of the atmosphere, the amount of solar energy the Earth receives, and how the resultant planetary heat circulates around the planet. Through the lens of geologic history, many climatic reconstructions depend on indirectly determining the greenhouse gas concentration of the atmosphere and modeling heat transport via the role that past continental configurations played in moving heat from the equator to the poles via the ocean, for example.

Some significant mysteries still persist in the paleoclimate world, including a few major discrepancies between what the proxy measures of greenhouse gas concentration would predict for heat distribution and what seems possible via models. The early Eocene Period about 50 million years ago is an example of this, where the fossil and paleoclimate record indicates very warm conditions at the poles, but the models cannot match those temperatures, even though they perform well during other periods (https://royalsocietypublishing.org/doi/10.1098/rsta.2013.0123). Additionally, the complex interplay between solar insolation, surface albedo, ice, and equator to pole heat transport makes our models of climatic transitions challenging, particularly since significant ice cover began accumulating on the planet a few tens of millions of years ago.

In other words, the world still needs paleoclimatologists. Their research has the potential to examine how the past Earth actually changed as a function of ice cover, or greenhouse gas concentrations, to better model how our future world will respond to human-driven climate change. Earth systems have a complexity that can only be crudely translated into the mathematical equations and assumptions of physics that inform climate models. This results in some degree of subtle uncertainty about the timeline of ice sheet collapse or aridification intensification, but for the most part, humans are hitting our climate system with such a huge hammer of carbon dioxide that these subtleties become almost insignificant, whereas the largest variable—how much greenhouse gas emissions from human activities will be in 2030 or 2050—drives the whole climate show.

Did you know that?

There is a difference between weather and climate

We are all familiar with weather, which is a combination of the temperature, rainfall, humidity, wind strength, etc. that we experienced yesterday, or are experiencing today, or might experience tomorrow depending on how good the forecasts are. Weather is very important to our day-to-day life, and we really depend on knowing about weather to plan our day, or our tomorrow, or our weekend. Hence, our friendly local weatherman, who knows quite a bit about weather patterns and meteorology (the study of weather patterns), typically has little or no training in climate science. Climate is the average of weather conditions in a region over about a decade time scale. So, here in the Midwest, climate would tell us that it is likely to be cold in the winter and hot in the summer, and even give an indication of how cold, or how warm, it usually is. Climate would say that you should usually be wearing a coat in Indiana in February and shorts in July. Does that axiom always hold true? Of course not, because weather gets in the way, sometimes sending a blast of warm air in winter. One memorable example was in the festive week-long lead up to the 2012 Super Bowl in Indianapolis, Indiana, when temperatures shot up to the lower 60s Fahrenheit...in February! Shorts and t-shirts weather—but not typical for the climate in Indiana winter. Two years later, Indiana and most of the upper Midwest and east coast of the United States were in the grips of a long winter of frigid temperatures and massive snowfalls. Also not typical considering climate records, but it certainly was the weather state of the time. So, climate defines what you expect to wear tomorrow, or next month, but weather dictates what you will actually wear. When Oklahoma Senator James Inhofe in February 2015 infamously held up a snowball on the Senate Floor to repudiate global warming, it would mean one thing had it been July, and another given that it was February. Obviously, this was theatrics coming from, perhaps not ironically, the Chair of the Senate Committee on Environment and Public Works, and not informed by obvious facts about weather and climate.

The Carrington Event was a massive Solar storm that disrupted global communications

We all realize how powerful the Sun is, but did you know how erratic it can be? On the evenings of September 1–2 in 1859, auroras could be seen across the skies, even close to the equator. In the Northern Hemisphere, we know these as the Northern Lights, or Aurora Borealis (in the Southern Hemisphere, Aurora Australis), and typically only glimpse them in the high latitudes on relatively rare occasions. These auroras are caused by interactions between material ejected from the Sun and our upper atmosphere. But on those evenings in September, the lights were strong enough to read a newspaper by at night in the Northeastern United States. The strange auroras were the result of a massive Coronal Mass Ejection (geomagnetic storm) coming from the Sun, which made a direct hit on the Earth. The effect was so strong that it knocked out telegraph lines across Europe and North America, although in some cases the telegraph lines were electrified enough by the charged particles that telegraph operators were able to send telegram, even after the power was disconnected. Since then, the Earth has experienced more of these, including most recently in 1989 when an event knocked out power all over Quebec, Canada. We narrowly dodged another such event in 2012, which missed the Earth—had it made a direct hit like the Carrington Event, however, it would have surely damaged or destroyed the vast array of satellites which are the vital telecommunication and observational links that we all rely on for data transfer, banking, maps. Imagine what it would have done to cell phone networks? Unlike sunlight, which travels the vast distance from the Sun to the Earth in 8 min, these Solar Storms typically take several days to reach the Earth as they are comprised of particles rather than photons. How do you think the Earth will respond with only 2 days advance warning of an event that may miss us like in 2012, or hit us full on like in September 1859?

The measurement of carbon dioxide in Hawaii has been a father-son project since 1958

Charles Keeling, a Professor at Scripps Institution of Oceanography at UC San Diego, began taking measurements of carbon dioxide high atop the volcanic peak of Mauna Loa on the big island of Hawaii. This site was chosen for its remote location and lack of nearby vegetation (which might affect the carbon dioxide measurements). This endeavor was reasonably expensive, and in the early years, funding shortfalls risked the continuous operation of this monitor. Once it began proving its worth, and yielding surprising data about not just the biosphere but the accumulation of carbon dioxide in the atmosphere, the value of continuous monitoring became apparent (https://www.science.org/doi/10.1126/science.1156761). The graph of the increase in carbon dioxide was named the Keeling Curve, and is an iconic symbol of human influences on the planet. Dr. Naomi Oreskes of Harvard University called the Keeling Curve one of the most significant works of the 20th century, alerting humanity to the impacts of human activities in a dispassionate, nonpolitical way. Charles died in 2005, but by then, his son Ralph also a Professor at Scripps Institution of Oceanography, took over operations at Mauna Loa. Its first measurement in 1958 recorded a carbon dioxide concentration of about 315 parts per million. Now 60+ years later, the value stands at 418 parts per million, and rises every year as carbon dioxide continues to be emitted by humans.

Other great discoveries in science have also been family affairs. The Leakeys, husband and wife team Louis and Mary, made seminal contributions to our understanding of early hominid history via various archeological digs in Africa, a tradition that passed on to their son Richard and his wife Maeve. You will meet the Alvarezes in a chapter on Mass Extinctions, but basically this was a father-son team of Luis and Walter who came up with the concept that a major asteroid collision killed off the dinosaurs about 65 million years ago. Luis' grandfather was a famous physician and Luis himself won a Nobel Prize in physics before conceiving of the asteroid collision, and thus the science ran deep in the Alvarez family!

CHAPTER THREE

Oxygen accumulation and the first major life–climate interactions

Introduction

It took life a billion years to make any difference at all to our planet. To an alien observer watching our beautiful blue and white and brown orb swinging around the Sun, it was definitely a boring program. Surely, the newly formed tectonic plates were moving around, occasionally smashing into each other and making profound mountain ranges or sinking into the Earth's interior and sparking massive chains of volcanoes. The planet continued spinning around its axis, bringing day and night, but at a rate of every 18 h instead of 24 h—the gravitational clutch of the Moon has been continuously slowing our orbit. Clouds formed from the ocean, and rain fell on the landscape, albeit one completely devoid of any life whatsoever. Waves crashed on lifeless beaches, and wind blew over arid plains. Gasses in our atmosphere remained comprised of nitrogen and carbon dioxide and methane, with little apparent change.

Even if our alien observer had a visit to our ocean depths, it would be a relatively unenlightening one. It would discover life, of course, but the simplest forms of unicellular bacteria without any discernible internal structure, doing the simplest kind of work—gaining a modicum of energy from chemically altering various compounds and producing as a byproduct oxygen gas, which was instantly sucked up by the abundant reduced iron molecules dissolved in the ocean or present as particles. But a billion years is a long time, and microbes are persistent—eventually, they overcame the Earth's capacity to absorb all their waste oxygen ... and plunged the Earth into its first of several massive Ice Ages. Our alien observers would certainly have noticed that and would have been interested in watching the "Earth Channel" after a couple billion years of otherwise boring programming!

This step-change in global Earth geochemistry, called the "Great Oxidation Event" or GOE, marked a key transition in our planet's dynamic environmental and ecological history, with a planet going from having no atmospheric oxygen to one that permanently has oxygen. Not much of it

Climate Change and Life
ISBN: 978-0-12-822568-4
https://doi.org/10.1016/B978-0-12-822568-4.00008-0

for a significant time after the GOE, to be sure, but enough to permanently change the ecological structure of organismal biology in the oceans, setting the stage for the eventual evolution of oxygen-hungry animals. As with all major transitions, the GOE carved its signature in the geologic record, and even changed the trajectory of the US and global economy in the early mid-20th century, in a roundabout way at least. Because it was so long ago, the exact nature and timing of all the dynamic changes during this time are a bit blurry, but stunning geologic examples of this transition period exist on many older, more stable continental interiors on various continents, providing convincing evidence of the sequence of events during this interval. In this chapter, we explore what was going on with geochemical and biological systems before, during, and after the GOE, and will discuss microbial metabolism, marine cycling, US industrial history, and the dynamics of our first Snowball Earth.

The simple but fascinating world of microbes

On the modern Earth, microbes are all around us. They are in soil, in the water, in deep cracks of the Earth, in boiling alkaline cauldrons of Yellowstone, in the darkest depths of the ocean, in the air, and, of course, on and in us. To identify the presence, abundance, and type of microbe in a sample, biologists typically take a swab, wipe it over a surface or dip it into a liquid, smear it onto a plate with different kinds of nutrients to make it grow superfast, and then examine the "cultured" plates over time. It is actually not hard work at all—other than some ultrasterilized surfaces in specially filtered rooms, everywhere you wipe you will find something. Literally everywhere. Our planet is an explosion of bacteria, and we as humans are vastly outnumbered.

There are about one trillion different species of microbes (Earth May Be Home to a Trillion Species of Microbes, 2016), and 99.999% of them have yet to be discovered (How Many Bacteria Live on Earth? 2017)! These one trillion species comprise a total of about 5,000,000,000,000,000.000,000,000,000,000 individuals, so we humans are outnumbered about 21:1. Technically, microbes are a blanket classification for a number of organisms that are simply microscopic (hence the term) including bacteria, virus, fungi, algae, lichen, and more. For the purposes of this chapter, we narrow in on bacteria, as they played the biggest role in early Earth history and continue to this day. Some bacteria do what land plants do—harness the energy of the sun through photosynthesis to make sugar, which they then metabolize for energy. The evolution of this photosynthetic pathway will play a part in our understanding of the GOE, but so will an awareness of other ways that microbes make a living. Some harness the potential energy stored in certain chemical reactions to produce sugar, which they then metabolize—certain types of these were the first organisms on the planet, and versions of them persist in certain environments to this day.

It is impossible to definitely document metabolic shifts in organisms during the GOE simply due to the limitations of the fossil record. But some researchers have applied a space-for-time substitution approach to infer core metabolic functions during this transition using microbial mats that exist in extreme environments today—like in the Salar de Llamará, one of the many "salares" or salt flats that occur in the Atacama Desert, Chile (Metabolic Evolution in Ancient Microbial Ecosystems, 2018). This is one of the driest locations on the planet, with extremely high daily temperature variations and high ultraviolet radiation owing to its high average altitude of about 3000 m above sea level. By measuring spatial relationships and relative abundances of metabolic genes in these maps, researchers hypothesize that although non-oxygen—producing photosynthesis was present before the GOE, it diversified in different microbes and expanded in parallel to cyanobacterial oxygenic photosynthesis. Second, the carbon fixation pathway shifted permanently to the modern Calvin cycle pathway of converting carbon dioxide to sugars after the GOE.

At the end of the day, for all of these differences of what they eat, and where they live, bacteria all share some basic attributes. They are almost all microscopic, typically ranging from 0.5 to 2 microns. They do not have internal structures, a characteristic that places them in the category of "prokaryotes" as opposed to organisms with complex internal structures (with strange names like mitochondrion and Golgi bodies and endoplasmic reticulum), which are called "eukaryotes." They have DNA, but only one strand, and not the famous "double helix" sported by all other organisms. And they are all autotrophs, meaning that they can produce their own food using substances in their environment like light or chemicals. Three basic classes of bacteria exist today: eubacteria, which are also called "true" bacteria and are abundant everywhere, the tough archaea, and the complex eukaryotes described above (Cooper, 2000). Interestingly, these three groups evolved from some common prokaryotic organism very early in Earth's history, such that they all co-existed by about 2.7 billion years ago, soon before the GOE.

The greatest polluter of all time—introducing the cyanobacteria

Cyanobacteria are the Earth's heroes, and villains. Cyanobacteria evolved about 2.7 billion years ago, and came on the scene with a novel innovation—they could harness the power of sunlight to make sugars, and in the process released waste oxygen. This does not seem that novel now, where we are surrounded by this basic photosynthetic mechanism, but it was 2.7 billion years ago. Before the evolution of cyanobacteria, all organisms metabolized via chemosynthesis, or converted chemical compounds to make sugars. This is a fine way to make a living and all, but the evolution of the cyanobacteria is akin to the advent of mass production—chemosynthesis is a relatively slow process, whereas cyanobacterial photosynthesis was like the Ford Assembly Line, kicking out thousands of cars per day. Like all organisms that learn to do something faster or better, cyanobacteria soon dominated the bacterial world. But much like the burning of gasoline and emissions of carbon out of tailpipes has been the unregulated global cost of using fossil fuels, cyanobacteria were similarly unregulated, spewing vast amounts of oxygen into the ocean depths. There it was rapidly whisked away by the formation of rust compounds from the abundant iron all around (Fig. 3.1).

Figure 3.1 ***Cyanobacterial Bloom.*** A bloom of cyanobacteria on a modern lake. Although different in some functionality, these modern cyanobacterias hearken back to the first oxygen-producing photosynthetic organisms on the planet, which evolved and proliferated shortly before 2.3 billion years ago. *Licensed under Education License, Indiana University.*

Early photosynthetic organisms used enzymes that captured ultraviolet bands of sunlight and, through a conversion involving sulfur, produced energy. The organisms are termed purple sulfur bacteria, and still exist to this day in unique lake settings where these depths are devoid of oxygen and sunlight is able to penetrate down to these deeper waters to fuel anaerobic photosynthesis. The early versions of purple sulfur bacteria thrived on abundant and deadly (to us at least) hydrogen sulfide, and produced solid sulfur minerals as a byproduct. These sulfur minerals would further break down into sulfuric acid—one can easily imagine this foul early planet that smelled of rotten eggs and whose water would burn our skins. But soon, another bacterial player came on the scene, that literally changed everything—a phototroph that feeds on water and carbon dioxide and belches out oxygen. The nearest equivalent to early cyanobacterial colonies or mats are thought to be most like modern stromatolites, pillar-like colonies of cyanobacteria that are found in only a few isolated environments on Earth—typically quite saline and hot coastal environments that experience regular tidal variations (Fig. 3.2).

Cyanobacteria are so-named because of their own unique sunlight-trapping enzyme that they developed, one which captures sunlight in the more abundant visible bands of red light. This is why most of them appear green—the reds are subtracted from the wavelength equations, leaving cooler colors that are reflected to our eyes. The unique enzyme that they created to capture sunlight in the visible red band is, of course, chlorophyll. Nowadays, we sprinkle it in our protein shakes for that extra green kick, but this enzyme, once "invented" by early cyanobacteria, went on to control how nearly every phototroph makes its living, from green algae to sunflowers to cacti to oak trees. Indeed, if you look at Earth from a satellite view, our planet reveals chlorophyll in the vast areas of the ocean and tropical rainforests that are green from space. Pretty amazing to consider that one enzyme, invented by cyanobacteria 2.4 billion years ago (Cardona et al., 2015), would go on to change the color of our entire planet!

Iron formations mark the oxygen transition

In 1844, William Burt was exploring the southern shore of Lake Superior in the State of Michigan (USA) for potentially valuable resources, and was dumbfounded by an errant compass reading. As he and his expedition traveled north next to a nearby outcrop of rock, his compass did not follow suit—instead of pointing straight ahead, it kept locked onto the rock

Figure 3.2 ***Stromalite Deposits in Shark's Bay, Australia.*** These pinnacles are cyanobacterial mats that cement together sand along the shoreline. Stromalites are a rare modern vestige of the vast early cyanobacterial mats that contributed the planet's first free oxygen gas just before 2.3 billion years ago. *Image courtesy: Jody Webster.*

outcrop. What Mr. Burt found in that outcrop shaped the industrial trajectory of the United States, and shaped our understanding of the ancient world (Fig. 3.3).

Termed "Banded Iron Formations," or BIFs for short, the deposits found by Mr. Burt are unique rock formations that do not form today. This is unusual, in that geological processes and particularly the deposition of sediments in rivers and the ocean follow certain physical rules. Those rules have not changed, and indeed were first laid out by the Scottish farmer and naturalist James Hutton in the mid-1700s. The context of Hutton's observations of geologic time was profound—this was at a time before Charles

Figure 3.3 ***Large Boulder of Ironstone Formed During the Great Oxidation Event.*** This ironstone boulder is a portion of a Banded Iron Formation in northern Michigan, USA. It represents not only the first major production of oxygen gas on the planet by oxygenic photosynthetic cyanobacteria, but also the major ore resource that built the United States of America into a major steel-producing nation. This original ironstone, evidence of extreme climate change 2.3 billion years ago, ironically is a marker of another climate tumult—the large amount of energy needed to produce steel from iron ore was largely sourced from the burning of coal in large blast furnaces, which produces massive amounts of carbon dioxide that now influences human-produced climate change. *Licensed under Education License, Indiana University.*

Darwin explored evolution and the necessarily long period of time required for new species to form, and to go extinct. The ruling paradigm defining the length of history was based on biblical generation counting, and compressed all of Earth's history to a narrow time window, starting around six thousand years ago. Like Darwin, though, Hutton was a slow and steady observer not of what was written in the scripture but rather what was written in the rock record. Also, he was in a unique setting to do so, given the rock outcrops along the coast of southeastern Scotland where Hutton tramped in the field reflected a long sequence of sedimentary events that were interrupted by intervals of missing material, periods of uplift and erosion, and various geological events that presented as time markers of a sort. From these observations came one of Hutton's profound intellectual contributions, and marked the birth of modern geology—the Theory of Uniformitarianism. This is an overly complex term that simply states that physical geological processes occurred in the past at similar rates as they do in the present. So the slow deposition of sediments in a lake or the ocean, and the bit-by-bit erosion of a hillside in Scotland observed by Hutton at that time, must be matched with the actual geologic evidence in front of him—namely, the vast, thick sequences of sandstones and shales and the dramatic erosional hiatuses in the hillsides. Hutton surmised that the Earth had a history far longer than 6000 years, a concept that Darwin latched on to interpret his observations of selective adaptation.

Back to BIFs. These singular deposits seem to fly in the face of Uniformitarianism, and thus could be explained by only one mechanism—transitions in Earth's chemical cycles (and not its physical ones) that no longer exist today. In the case of BIFs, the general agreement is that this transition was provided by the rapid production of free oxygen in the oceans from early photosynthetic organisms—yep, those cyanobacteria again! In this case, deeper anoxic (containing no free oxygen) ocean waters rich in unoxidized iron produced by volcanic reactions on the seafloor (Fig. 3.4)

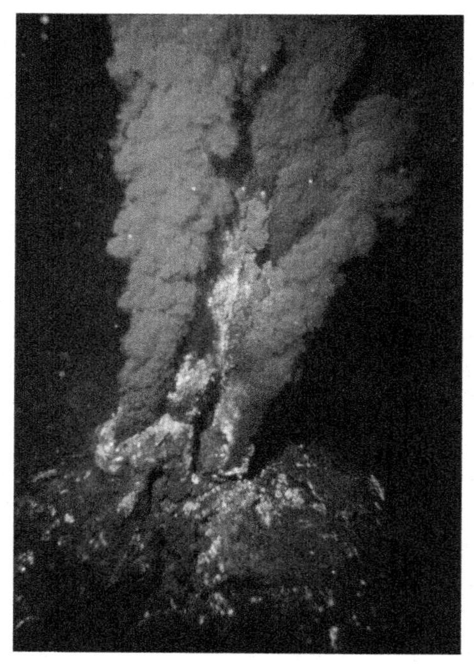

Figure 3.4 ***Mid-Ocean Ridge Vent Emitting Iron Particles into the Deep Sea.*** The Earth is constantly recycling elements, including iron. Here black iron particles emitted on the sea floor from volcanic vents are produced by interactions between seawater and the magma below the surface. A similar process would have occurred on the early Earth, providing a substantial supply of unoxidized and dissolved iron—not particulate iron as in the modern oxygenated ocean example shown here.

upwelled toward the surface ocean where photosynthetic, oxygen-producing cyanobacteria were now thriving. The dissolved iron was rapidly oxidized and became particulate mineral iron, depositing onto the upper ocean seabed in reddish sedimentary layers. This process was intermittent, as the BIFs are indeed banded, with red or black layers, representing oxidized iron, interrupted by gray shale or sand layers likely representing episodic pauses in deep ocean upwelling, consistent with similar pauses that we see today and represented by layered sedimentary compositions—with the key difference that these layers do not include precipitated iron. Interestingly, these anoxic upwelling events were actually critical to the cyanobacterial ecosystems at the time. They served to consume the oxygen waste product via iron oxidation that otherwise would choke out early cyanobacteria, which had not yet evolved certain enzymes, such as superoxide dismutase, which later allowed them to live in an oxidized environment (Fig. 3.5).

William Burt's discovery of BIFs in the upper peninsula of Michigan was literally a gold mine (or more fittingly, an iron mine) for the early industrialists of the United States of America. This is simply because, during a brief period of time, all the unoxidized iron in the oceans at that time was

Figure 3.5 ***Banded Iron Formation (BIF) in Cross-Section.*** Red (*light gray in print version*) and gray stripes on a sample of a BIF reveal alternating periods of oxygen-rich (red [*light gray in print version*] oxides) and oxygen-poor (gray sandstone) conditions during the transitional period of the Great Oxidation Event 2.3 billion years ago. *Licensed under Education License, Indiana University.*

scrubbed out and deposited on continental margins around the world, including the one that was present about 2.3 billion years ago where Michigan is now. These Michigan BIFs became vast iron mines, and their location right on the shores of the Great Lakes was one of those great geological coincidences. One of the biggest challenges with hauling heavy ores across the country at that time was limited rail service, but there is no better media for carrying heavy loads at relatively low costs than water—in this case, via barges that carried the ore southward from Lake Superior to the southern shores of Lake Michigan where they met their fiery match in the vast smelting furnaces that began lining the shoreline. There huge furnaces were fueled by tons of coal that were delivered by rail from coalmines just a few hundred kilometers to the south, in Indiana and Illinois. The BIF-derived iron ore was 2.3 billion years old, whereas the relatively nearby coal deposits formed only about 0.3 billion years. In both cases, they were formed in shallow waters of an ancient sea, but completely separated by a vast gulf of geologic time. Regardless, this region just east of Chicago, Illinois became the greatest producer of iron and steel in the world, and it is no coincidence that the birthplace of industrial mass production and automobiles was only a short distance away in Detroit, Michigan. Thus, geology not only teaches volumes about deep time but also about the resources that drive the arc of human history and development.

And then ... the planet froze over

It seems a nonsequitur to link rising oxygen concentrations and BIFs with the first massive global icehouse event (but not the last, as you will read elsewhere), but these are indeed linked, and the global impacts were profound. Dissolved oxygen gas in the oceans and bubbling up to give our atmosphere its first breath of oxygen sounds great to us of course. As noted above, this was not so great for the organisms that now had to hide out in the sparse environments without oxygen. But that newly liberated oxygen also had another ecologically destructive trick up its sleeves—it literally stripped methane out of the atmosphere. This seems a very unlikely one-two punch for Earth's early ecosystems—first these early photosynthetic organisms poisoning from oxygen itself, extinguishing entire ecosystems and driving formerly dominant organisms into dark, deep environments, and then killing even more by causing a global freeze-over with ice so thick that very little sunlight penetrated into the ocean waters below. But when the geologic record shows that this was what happened, first with the BIFs characteristic of oxygen accumulation and then global glacial deposits consistent with ice sheets existing all the way to the equator, an explanation needs to be made (Fig. 3.6).

Based on the Sun's weak early output, a global ice-over at that time was actually not at all inconsistent with the lower solar radiation at that time. And just like today, without the presence of any greenhouse gasses in the atmosphere, the Earth would be well below freezing 2.3 billion years ago. But at this time, the Earth had a relatively thick greenhouse gas cover, particularly the powerful greenhouse gas methane. This early Earth atmosphere was a reducing one before oxygen. But not only were the abundant iron minerals in a reduced form (not oxidized or rusted), but so was much of its atmospheric carbon. To briefly explain, the carbon in carbon dioxide has a charge of +4, which counteracts the two attached oxygen atoms which have charges of −2 apiece. In contrast, the carbon in methane has a charge of −4, to counteract the four attached hydrogen atoms which have charges of +1 apiece. Thus, the carbon in carbon dioxide is oxidized (low in negative charges), while that in methane is reduced (high in negative charges). For the curious, those charge differences have to do with how many electrons, which have a negative charge, are in the outer atomic shell of carbon

Figure 3.6 ***Thick Ice Cover that Would Have Been Typical of Conditions During the First Snowball Earth.*** The surface of the Earth completely froze over during its first "Snowball" event about 2.3 billion years ago. *Licensed under Education License, Indiana University.*

in each molecule. The chemistry of carbon is stable in each molecule, but if the environment changes, much like iron, the carbon can change to suit the new environment. And 2.3 billion years ago, this environmental change came from the newly liberated oxygen in both cases.

As noted in Chapter 2, methane is a very powerful greenhouse gas—a little goes a long way. Depending on how it is calculated, 1 molecule of methane is worth 26 molecules of carbon dioxide with respect to its ability to warm the planet. Our early Earth, bathing in a methane-rich atmosphere, was thus able to offset the feeble early Sun and less-than-ideal orbital position on the outer edge of the Solar System's Habitability Zone to stay above freezing. Of course, this condition was largely relying upon the greenhouse warming provided by methane, which in turn was relying on having no oxidants in the atmosphere. At 2.3 billion years ago, however, those first whiffs of free atmospheric oxygen were just a shot across the bow, and likely were quickly stripped out of the atmosphere by oxidizing just a fraction of the methane, doing little to impact global temperatures. But as the ocean's oxygen-absorbing capacity was overwhelmed by profligate phototrophic production of oxygen and complete oxidation of iron, oxygen gas began venting into the atmosphere at a rate that overwhelmed the absorptive capacity of the atmosphere. This ultimately results into the oxidation of nearly all the atmospheric methane (some methane likely continued to be produced by methanogenic bacteria, much like today). It is true that the methane, once oxidized to carbon dioxide, retained some capacity to absorb heat as a greenhouse gas, but at a significantly reduced scale. So, a planet poised on the edge of a deep freeze because of conditions in the solar system itself finally and quickly succumbed, turning into a shiny orb of ice. Our ancient alien observers would see a planet that had been rather dim, absorbing a significant amount of incoming solar radiation in its oceans and on land, light up in their telescopes.

The concept of albedo was introduced in the Chapter 2, and would have played a role in our first global "Snowball Earth" event. As methane was stripped out of the atmosphere and the poles lost some of their warmth from greenhouse forcing, they would have started forming ice, which would reflect more incoming radiation back out to space without warming the planet, which would allow the ice to expand, which would ... you get the picture. But at this point, stable states and nonlinear responses are critical to understanding why the Snowball Earth was not permanent, either then at 2.3 billion years ago or the several successive Snowball events that came later. Basically, the Earth was nudged into an inertial valley of sorts, achieving a steady state. To get out of that valley, a significant energy push needs to be made to roll it back up the slope into its previous ice free state (Fig. 3.7).

There exists some debate about the absolute timing of the Great Oxidation Event, which led to the global glaciation of 2.3 billion years ago The glaciation interval is called the Huronian glaciation, and roughly brackets and age range of 2.45—2.22 billion years ago. This question of timing was the topic of a rock core drilling expedition that occurred in northwestern Russia, which targeted sedimentary rocks that were formerly in the ocean between 2.5 and 2.43 billion years ago (Warkea et al., 2020). Using techniques that analyze sulfur isotopes in the rocks (sulfur isotopes are very sensitive to changes in free oxygen concentration in water), researchers found that rocks earlier than 2.5 billion years ago definitely do not indicate the presence of oxygen, while those afterward do. Thus, the Great Oxidation Event seemed to have occurred in a narrow time window (geologically speaking at least!) between 2.5 and 2.43 billion years ago. This timing is critical, because it is consistent with the model above of the release of free oxygen into the oceans and atmosphere stripping out planet-warming methane and triggering a global freeze over.

But back to the energy well that the Earth found itself in during the grips of the global freeze-over—how did the Earth pull itself out of the well? For

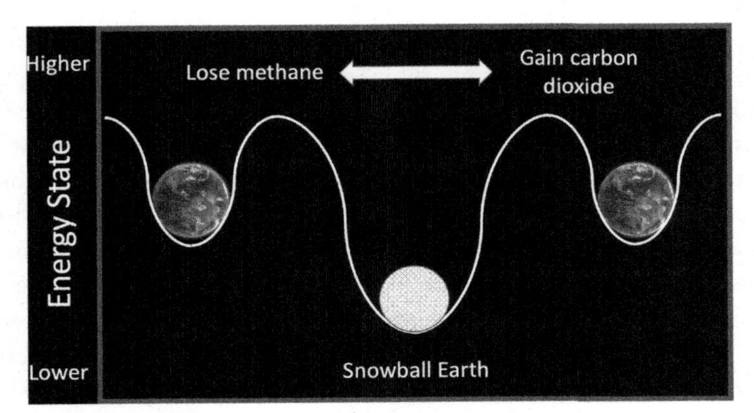

Figure 3.7 ***Energy Wells and Snowball Earth.*** Earth's early climate was defined by different energy states, with a relatively warm climate existing when the Earth had high greenhouse gas warming from methane, a frozen, Snowball state the newly introduced oxygen stripped out methane from the atmosphere, and a return to a warm state when enough carbon dioxide had accumulated in the atmosphere to enhance the greenhouse effect.

that, we cannot rely on any external climate forcing, as the weak early Sun did not suddenly warm up—this process is quite gradual. One intriguing, and quite dramatic, hypothesis for the warm up is again linked to greenhouse gasses, but this time a passive one—water vapor. As discussed in Chapter 2, water vapor is also a greenhouse gas, but typically is a passive one, in that if the Earth warms, the atmosphere is able to hold more water which in turn can enhance warming. The inverse is also true, of course. Where did the water vapor come from? One study of an impact crater in exotically named Yarrabubba, Western Australia (The Earth was once a snowball. an asteroid impact two billion years ago may have changed that, 2020) that dates to 2.23 billion years ago used the hydration state of minerals there to infer that the massive asteroid that caused this 70 km wide crater slammed into the ice-covered planet and instantly vaporized a tremendous amount of water (Erickson et al., 2020). This is equivalent to about 1% of the water vapor present in our relatively wet atmosphere, and when jetted into the dry air of Snowball Earth would have had massive and instantaneous warming effects. One potential weakness of this hypothesis has to do with how long water vapor could stay in the atmosphere. If the water injection stayed in the atmosphere for a significant amount of time (millions of years?), it might have "reset" Earth's early climate, and the planet might have been able to maintain ice-free conditions, or at least mostly ice-free, through the greenhouse actions of water vapor and carbon dioxide alone. But this process is not at all certain, and warrants other possibilities.

A competing hypothesis to the asteroid impact-water vapor scenario relied solely on carbon dioxide. Carbon dioxide is removed from the atmosphere through weathering reactions that occur in rocks. If conditions are ripe for a significant amount of global weathering, namely a wet and warm atmosphere that speeds the process of mineral dissolution, the rate of carbon dioxide removal from the atmosphere can be significant, leading to cooling. Eventually, a so-called negative feedback loop kicks in to slow and ultimately stop this process—as carbon dioxide concentrations go down, the planet gets colder and the atmosphere drier, leading also to reductions in water vapor which, combined with the lower temperatures, slow mineral weathering. But, the inverse of the above also occurs, namely that if the planet is iced over, mineral weathering is reduced to nearly zero as temperatures are low and there is little interaction between the atmosphere and the now ice-topped rocks layers below. Thus, there is little extraction of atmospheric carbon dioxide, and as volcanic sources continue to add this greenhouse gas to the atmosphere, perhaps via carbon dioxide "volcanoes" piercing through the ice sheets, the atmosphere warms and the ice eventually melts. Thus, much as occurs with recent Ice Ages, this model relies on simple variations in the carbon cycle to eventually thaw the planet and needs no extraterrestrial visitor, in the form of an asteroid, for the Earth to recover. We will see in Chapter 4 that global snowballs and recovery from them tend to invoke this carbon dioxide-only process, but with the important difference that by then, the Sun was getting close to its current radiation output and thus it would be "easier" for carbon dioxide alone to be the ice-recovery trigger.

One aspect of the period of recovery from the 2.3 billion year ago freeze-over is that the planet got pretty strange for the subsequent 0.25 billion years (250 million years), during the so-called Rhyacian Period. It seems that the planet was not over with massive glaciation events, and evidence exists for another four phases of glaciation during this time. These might not have been the global catastrophe that marked the first event, but might have represented a critical transition period after which the planet went for well over a billion years before the next major glaciation. This enigmatic period was marked by a significant lull in tectonic and volcanic history—a temporary slowdown of the Earth's internal dynamo that is not well understood. The rock record does reveal warm intervals between the four glaciations, marked by limestone deposition from warm, ice-free seas. These warm periods might have been related to sporadic volcanic activity amidst the relative lull, but might have instead been driven by biological responses, which brings us to a critical question—what were ecosystems doing through this dynamic, and environmentally traumatic, time?

Ecosystems in turmoil

The Great Oxidation Event is considered one of the major step-changes in Earth's global environment—before no oxygen, after always oxygen. To be clear, the "always oxygen" part of that does not mean that we could breathe in the post-GOE world. Although oxygen might have shot up to close to modern levels at one time during the transition, soon afterward, as environments stabilized, oxygen plateaued off at about 5%–20% of modern levels and stayed at those levels ... for over a billion years! This post-GOE "boring billion" interval is likely anything but boring, but one thing is clear—oxygen never declined to zero, or increased to near modern levels, during it. But even at relatively low levels by modern standards, the very presence of oxygen sparked one of the most critical evolutionary developments in Earth's history—the development of eukaryotic organisms, of which we are an extension.

Before introducing eukaryotes deeply, it is important to understand what limits life in the first place. Basically, what are the major "speed limits" that control all ecosystems on the planet, and thus the organisms that live in them, be they algae or avocados or aardvarks? We are all organisms of carbon, that simple element that is present in myriad forms, such as limestone and teeth and oil, and is also vital to all organisms. For cyanobacteria, their entire way of living involves capturing the power of sunlight in a web of carbon. As noted earlier, they use chlorophyll to temporarily trap solar radiation, and then tap into that energy to "crack" water and combine it with carbon dioxide gas to produce sugar, which they happily consume. The amazing molecular flexibility of carbon makes this possible through its penchant to form molecular rings, the basis for so-called organic molecules that are in all other organisms. At the other end of the carbon consumption chain might be the aardvark, which consumes ants and termites loaded with carbon sugars that they extracted from their own diets but which now fuel the aardvark to do whatever it is that aardvarks do. Ultimately, we all revel in the ability of photosynthesis to trap sunlight into sugars, and the fact that we have oxygen around to burn that sugar. And this latter requirement—that oxygen is required to easily and rapidly release the energy stored in carbon sugars—is why the evolutionary step-change occurred after the GOE.

But other ingredients are necessary for life. We need calcium for teeth and bone development and vitamin A for our vision and fat for metabolism and protein for muscle development and the full spectrum of macro- and micronutrients that appear in the nutritional facts labels of our breakfast cereal. These nutritional needs are because we are relatively complex multicellular animals, and indeed share many of these same nutritional needs with other members of our mammalian extended family. Cyanobacteria, on the other end of the nutritional needs spectrum, require only three major ingredients and a host of very minor ingredients. They have no need to consume vitamin A (ever seen an algae with eyes? I didn't think so) or fat or protein—more sophisticated plant organisms do need some of that, but they produce it themselves, much like we produce our own vitamin D through sunlight interactions in our skin. But, along with carbon, cyanobacteria need a ton of nitrogen, and a little less phosphorus. This need for the three main ingredients, carbon, nitrogen, and phosphorus, is shared with every plant that you see around you, be it algae or an aster flower or an ash tree. The multicellular aster and ash have a lot more secondary nutritional requirements as they have to produce biological structures such as waxes for leaf function and lignins to grow tall.

An organism's access to carbon seems not to be a huge problem, as carbon is present as carbon dioxide, methane, dissolved carbonate ions, etc. More challenging for modern ecosystems is nitrogen, a critical element for biosynthesis. There is plentiful nitrogen on our planet—for example, our atmosphere is 80% nitrogen—but it is hard for organisms to get to it. The atmospheric nitrogen gas is bound up in a stable and hard-to-break molecular headlock between two nitrogen atoms. Some of the nitrogen that is released from the weathering of rocks and by dissolution of nitrogen into precipitation does make it to the oceans to fuel marine ecosystems, but organismal demand is high. An interesting symbiotic system is present on land and in the sea to alleviate the potential nitrogen shortfall—namely, nitrogen fixation. On land, this is a common occurrence and is a process accelerated by certain types of plants, such as legumes. The roots of legumes are colonized by rhizobium bacteria, which have a wonderful symbiotic relationship with the plants themselves—they consume a little sugar that the plants produce during normal photosynthesis, and in return convert atmospheric nitrogen gas into a useable nitrate ion for the plant. This occurs at sea as well, through another stand-alone bacterial process but with the same net results—breaking the powerful nitrogen–nitrogen bond and converting nitrogen gas into a nutrient. Both processes take some time. For example, an area might experience rapid algal growth after a particular injection of nitrate which consumes all the available nitrate—upon which the bloom collapses. But bacteria will restore the nitrate after several weeks, and the ecosystem can return to normal functioning.

So, carbon? Check! Nitrogen? Check! How about that other nutrient, phosphorus, which is not needed by organisms in nearly the same quantities as carbon and nitrogen but is still absolutely vital to their metabolism and cell structure? Here is where things get difficult for a cyanobacterium, for a host of reasons. First off, there simply is not much phosphorus on the planet. Rocks on Earth contain less than 0.1% phosphorus on average. Life may have had a luckier start on Mars, which appears to have twice as much phosphorus in its rocks. But of course we are stuck here, as was our hungry cyanobacteria 2.3 billion years ago. But the real problem is that phosphorus just does not have the variety that carbon and nitrogen do. There is no vast reservoir of phosphorus gas in the atmosphere or dissolved in the oceans for ecosystems to draw from. There is only the slow trickle of phosphorus weathered out of rocks and transported to the ocean. For these reasons, the writer and chemist Isaac Asimov, in a 1959 essay, dubbed phosphorus "life's bottleneck."

Here, another issue comes up that likely contributed to ecosystem stress during and shortly after the GOE. The phosphate ion absorbs rapidly to iron oxide minerals (i.e., rust), taking it "out of commission" for ecosystems. In the modern ocean, this can be a temporary removal, as some areas of the ocean experience low oxygen conditions, particularly near coastlines, and oxygen can be depleted in ocean sediments through the process of respiration. Both can lead to a change in iron from a particulate iron oxide back to a dissolved ion, releasing the adsorbed phosphate ion in the process and returning it to the environment for reuse by photosynthetic organism. This process can occur endlessly, and in fact can fuel large algae blooms in a self-feeding way—phosphorus enters systems, algae bloom and eventually drop to the lower water or sediments, phosphate is released back to the water and brought back up to the surface by circulation, fueling another algae bloom, etc. This process is called the "iron–phosphate pump," and functions in most lower-oxygen settings in ocean water, lake water, and especially in sediments, where oxygen diffusion cannot easily keep up with oxygen utilization by animals. Of course in early Earth history, devoid of animal life, the main action occurred in the water column itself, with low oxygen driven simply by the relatively low production of oxygen by cyanobacteria at that time (Poulton, 2017). But even with this iron–phosphate pump functioning in relatively high gear during and after the GOE, leading to moderate amounts of phosphate being internally recycled in the ocean, the other issue continued to persist of low rates of phosphate delivery to the ocean in general (Filippelli, 2011), meaning that a preoxygen ocean was not necessarily entirely awash in phosphate.

But consider the transition from a world with very little free oxygen to one where oxygen is always present, albeit at lower levels than today. On the "no free oxygen" Earth, phosphate ions would be mostly free of their iron oxide captors, because there were very few iron oxides. Hence, the phosphate would build up to relatively high levels, until other mineral saturation reactions would top them off at a certain level. Thus, early organisms did not face the level of phosphorus concerns as they do in modern ecosystems. But once plentiful oxygen came on the scene and iron started oxidizing at a great rate, the removal of free nutrient phosphate ion was accelerated (this trend was later reversed, perhaps helping give rise to more complex organisms and animals about 1.5 billion years later) (Filippelli, 2010). This phosphate limitation, coupled with the difficulty that early forms of photosynthetic cyanobacteria had in coping with their own waste products (i.e., oxygen), puts a great strain on these organisms, and probably marked one of the first mass extinctions that the Earth experienced. Why "probably?" Well, much of this history is inferred by conjecture, as early soft organisms like cyanobacteria left no fossil evidence of their existence (or extinction, for that matter)—only some vague impressions on ancient sediments that are spotty through space and time. Much of what we know about these dramatic shifts in nutrients and ecosystems comes from some scant geochemical evidence from rocks and a lot of conjecture. Nevertheless, geologists live by the credo that the "present is a window into our past" and this process of phosphate limitation upon the introduction of oxygen into an ecosystem occurs today.

Eventually, this oxygenation–glaciation cycle was broken about 2.2 billion years ago, as evidenced by an increase in the amount of organic carbon buried in marine sediments. In a geochemical balancing act, more organic carbon burial equates to higher amounts of oxygen in the atmosphere, which suggests that photosynthetic organisms were having a heyday (Gumsley et al., 2017). This tipping point occurred through volcanic activity in this period, which provided a new influx of nutrients to the oceans, and finally giving cyanobacteria everything they needed to thrive. Oxygen levels were high enough to permanently suppress methane's oversized influence on the climate, thus ending methane's rule as the greenhouse gas king and ushering in carbon dioxide as the main driver of atmospheric heating on the planet.

Evolution on a stabilized, oxygenated planet

During and shortly after the Snowball Earth episode, phosphorus was still being delivered to the oceans to some extent through the weathering of rocks on land, a process that likely accelerated as volcanically sourced carbon dioxide started accumulating in the atmosphere, offsetting the oxidative loss of methane and returning Earth to a greenhouse gas–rich environment. It should be clear by now that the GOE-Snowball Earth interval was not like a light switch, with a clear off-on instantaneous transition. Even if, through the lens of geologic time, it can certainly seem that this was a rapid transition. But the planet likely vacillated through a set of low oxygen–higher oxygen and higher phosphate–lower phosphate cycles before stabilizing into an entirely new planet, which brought an entirely new set of opportunities for organismal evolution through the power of oxygen.

Oxygen is highly efficient in biological energetic reactions. Indeed, it is efficient in many chemical reactions, supporting the "state change" shift in geochemical cycles invoked by geochemists at this time. In biological systems, oxygen provides much more free energy than do most metabolic pathways, and it was like a new source of power, or new frontiers, had opened up to biological evolution. First, although initially relatively toxic to bacteria surrounding the oxygen-producing cyanobacteria, oxygen exerted a selective pressure that drove the evolution of Archaea bacteria into the first eukaryotes (Gross & Bhattacharya, 2010). This evolution might have been mediated by the role that reactive oxygen species and resultant oxidative stress, plus other environmental stresses, had in driving Archaea to adapt and evolve into more complex organisms that had internal structures. One of these structures, the cellular energy engine called the mitochondria, gave organisms the energy to exploit more complex structures and thus adapt to and expand into a wider range of ecological niches.

And let's not forget sex! Yes, this is referring to bacterial sex, which first evolved at this time. This process for bacteria is termed meiosis, and refers to the equal splitting and replication of DNA as two cells split from one. This is the core of how organisms function now, and is necessary for DNA repair. Although early organisms likely had some form of internal DNA repair (necessary to stabilization of organismal integrity over time), once mitochondria came on the scene and began producing reactive oxygen species inside cells, the need for better repair mechanisms became acute. There is no more efficient mechanism for DNA repair and stabilization than meiosis, and additionally, this new way of functioning added various complicated cellular mechanisms that opened up myriad environmental possibilities for eukaryotes.

Summary

It might seem that all the various organisms in the world evolved through random happenstance and accident. But in a sense, the algae, the aster flower, the ash tree, and the ant-eating aardvark were the result of a long series of moderately predictable interactions between chemicals, climate, and selective pressure. The playing field was set by geologic processes such as plate tectonics, chemical processes such as elemental interactions, climatic processes such as the greenhouse effect, and biological processes such as evolution. None of these processes exist in isolation, and they all come together in dramatic fashion in certain intervals of Earth history. The Great Oxidation Event and subsequent global glaciation during the Huronian Event was one such interval.

It could be argued that it was the most important one in Earth history, without which the modern world of algae and asters and ashes and aardvarks would not exist. But it could be argued that it simply had to happen at some point. A planet like Earth with all the right ingredients would eventually form life. This life would eventually alight on oxygenic reactions and ultimately photosynthesis as the way to go. The planet would eventually rust as plate tectonics would not be fast enough to keep loading new unoxidized iron into the system. Oxygen would strip out methane and plunge the planet into a global ice age. Even this disaster could not persist, as ultimately continued production of carbon dioxide would help offset the loss of warming methane. Additionally, all of this new, highly reactive oxygen would force organisms to be resourceful and adaptive to avoid damage, and voilà—those early algae set the planet on the pathway to contain asters and ashes and aardvarks, and apes, and us, all at the same time!

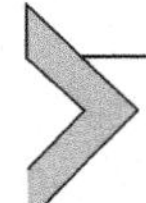

Did you know that?

Carbon and oxygen have a unique dance on our planet, but oxygen persists longer

The "dance" between carbon and oxygen on the planet has existed for as long as free oxygen first accumulated in the atmosphere. In simple terms, the dance is easy to learn—when oxygen is high, carbon dioxide is low, and vice versa. Consider what happens when carbon dioxide increases in the atmosphere. More carbon dioxide equals more organic carbon production on land and at sea by plants. A portion of that now-organic carbon is consumed by animals who pull oxygen from their environment in doing so. An oft-quoted concern of human-induced climate change is that all the carbon dioxide we are producing will result in more plant growth, more organic matter burial, and the related drawdown of oxygen. This concern is ill-founded.

A decline of oxygen from human activities that produce carbon dioxide is not likely because these two elements are unequal partners. The shear amount of oxygen in the atmosphere and ocean is huge compared to that of carbon dioxide. The atmosphere is comprised of 20% oxygen and only 0.042% of carbon dioxide. Thus, large swings in carbon dioxide concentrations barely nudge the amount of oxygen present. Another way to consider this is that the one molecule to one molecule relationship of carbon dioxide increase and oxygen decrease makes a lot bigger difference to the small pool of carbon dioxide molecules in the atmosphere/ocean system compared to the huge reservoir of oxygen. This results in significant stability of the oxygen pool in the atmosphere, even on longer time frames, which is critical given the very narrow range in oxygen concentrations that many animal species can survive within.

Animals can live in an oxygen-free world

A common assumption is that animals die when there is no oxygen available. Certainly we see this with fish kills caused by low oxygen waters and by the obvious example of drowning, but the assumption of oxygen-dependent animals is one forged by our own experiences and our very mitochondria. Mitochondria are the power houses of our cells, and the cells of nearly every other animal in existence today. These structures use oxygen to produce energy molecules in the cell, and likely evolved billions of years ago in an Earth environment that had free oxygen. Like much of life on the planet, the system evolved with the ingredients around it, and the oxygen conversion to energy molecules extremely efficient. But a new discovery in the depths of the Mediterranean Sea turns the assumption of oxygen requirements for animals on its head (https://www.nationalgeographic.com/animals/article/100416-oxygen-free-complex-animals-mediterranean#:~:text=Deep%20in%20the%20Mediterranean%2C%20scientists,known%20to%20live%20without%20oxygen.&text=April%2017%2C%202010-,Deep%20in%20the%20Mediterranean%2C%20scientists%20have%20discovered%20the%20first%20complex,known%20to%20live%20without%20oxygen).

These new animals, microscopic and jellyfish-like, were found surviving comfortable in the oxygen free waters of the deep Mediterranean Sea. What sets them apart is that their mitochondria function differently from typical animals, and can produce energy molecules without oxygen. Researchers conjecture that perhaps they are remnants of some of the more complex organisms that might have existed billions of years ago before free oxygen was present, a sort of living fossil. Or they might simply have evolved from oxygen-consuming animals to tolerate and then thrive in a niche where they have no other animal competitors. Either way, this finding certainly does expand our horizons about the range of conditions where complex life is found, both here on Earth and perhaps on oxygen-free planets circling other stars.

References

Cardona T, Murray J, Rutherford W. Origin and evolution of water oxidation before the last common ancestor of the cyanobacteria. Molecular Biology and Evolution 2015;32(5): 1310–28. https://doi.org/10.1093/molbev/msv024.

Cooper G. The cell: A molecular approach. Sinauer Associates; 2000. https://www.ncbi.nlm.nih.gov/books/NBK9839/.

Earth may be home to a trillion species of microbes. 2016. https://www.nytimes.com/2016/05/24/science/one-trillion-microbes-on-earth.html#:~:text=According%20to%20a%20new%20estimate,have%20yet%20to%20be%20discovered.

Erickson T, Kirkland C, Timms N, Cavosie A, Davison T. Precise radiometric age establishes Yarrabubbaensis, Western Australia, as Earth's oldest recognised meteorite impact structure. Nature Communications 2020;11. https://doi.org/10.1038/s41467-019-13985-7.

Filippelli G. Phosphorus and the gust of fresh air. Nature 2010;467:1052–3.

Filippelli G. Phosphate rock formation and marine phosphorus geochemistry: The deep time perspective. Chemosphere 2011. https://doi.org/10.1016/j.chemosphere.2011.02.019.

Gross J, Bhattacharya D. Uniting sex and eukaryote origins in an emerging oxygenic world. Biology Direct 2010;5(53). https://doi.org/10.1186/1745-6150-5-53.

Gumsley A, Chamberlain K, Bleeker W, Söderlund U, de Kock M, Larsson E, Bekker A. Timing and tempo of the great oxidation event. Proceedings of the National Academy of Sciences of the United States of America 2017;114(8):1811–6. https://doi.org/10.1073/pnas.1608824114.

How many bacteria live on Earth?. 2017. https://sciencing.com/how-many-bacteria-live-earth-4674401.html.

Metabolic evolution in ancient microbial ecosystems. 2018. https://natureecoevocommunity.nature.com/posts/39438-metabolic-evolution-in-ancient-microbial-ecosystems.

Poulton S. Early phosphorus redigested. Nature Geoscience 2017;10:75–6. https://www.nature.com/articles/ngeo2884.

The Earth Was Once a Snowball. An Asteroid Impact Two Billion Years Ago May Have Changed That. 2020. https://www.newsweek.com/snowball-earth-asteroid-impact-australia-1483219.

Warkea M, Di Rocco T, Zerkle A, Lepland A, Prave A, Martin A, Ueno Y, Condon D, Claire M. The great oxidation event preceded a paleoproterozoic "snowball Earth. Proceedings of the National Academy of Sciences of the United States of America 2020; 117(24):13314–20. https://doi.org/10.1073/pnas.2003090117.

CHAPTER FOUR

Snowball Earth and the most extreme climate states that the Earth has experienced

Introduction

Mapping the geology of south-central Australia in the late 1880s, geologists found a strange thing—a rock deposit that was formed by an ancient glacier. OK, this is not altogether strange given how many past glacial events have occurred on Earth, particularly since the beginning of the Pleistocene Ice Age intervals that began in the northern Hemisphere almost 3 million years ago. Indeed, if you travel anywhere north of Chicago, or Cologne, what you see around you is mostly the remains of the repetitive Ice Ages and their glacial assault on the landscape. But these cities are at a reasonably high latitude, where the impacts of orbitally-driven glaciation dominate. And the periods of ice sheet growth, advance, and decay are on the order of tens of thousands of years. What the geologists in Australia were scratching their heads about is that this area was near the equator about 700 million years ago, when the rocks that they were studying were deposited. And the area was at low elevation back then, not in the rare equatorial settings where you currently find localized glaciers, which are at extremely high elevation where the colder and drier air allows for ice formation. And to add to the mystery, this random finding in Australia was subsequently replicated in northern Namibia and the North American Cordillera. Collectively, these records also indicated that the equatorial ice sheets persisted not for tens of thousands of years, but rather for tens of millions of years.

A disconcerting picture was emerging about a planet that was at that time effectively frozen over—a so-called "Snowball Earth." Ice sheets over all the land surface, and even floating sea ice over nearly all of the ocean, with the possible exception of some open water in the equatorial oceans. Seen from outer space, our planet would have been transformed from the blue, brown green and white orb (Fig. 4.1) that typically reflects our distribution of ocean, land, and polar ice, to a bright, highly-reflective all-white planet (Fig. 4.2). We have some close analogs to what the Earth might have looked

Climate Change and Life
ISBN: 978-0-12-822568-4
https://doi.org/10.1016/B978-0-12-822568-4.00004-3

Figure 4.1 ***The Earth From Space.*** The blue, brown, green, and white planet that we call home, where water on the surface is in liquid, solid, and vapor phases.

Figure 4.2 ***A Snowball.*** The snowball sphere represents what the frozen surface of the Earth might have looked like during its Snowball Earth states.

like during its Snowball phase right here in our own solar system—the Jovian moon Europa and the Saturnian moon Enceladus. But these bodies are much smaller than the Earth, and much farther away from the Sun. Intriguingly, astrobiologists have their sights set on both of these moons in a quest to understand the evolution and persistence of life under a thick layer of ice. Should proposed expeditions to these moons be carried out, we may know more about how ice-covered bodies function and whether life might be possible on other icy planets in the universe.

More discussions about Europa as an exciting potential evolutionary analog will come later in this chapter, but the very possibility of a long Snowball Earth scenario raises a host of critical questions: How did we come to this extreme state after the previous billion years of relative planetary boredom? How did it persist so long, even as various Earth system feedbacks were functioning to bring about a thaw? How did life survive through this long freeze, and indeed, how and why did it suddenly explode in a fit of evolutionary creativity that brought with it the true beginning of an animal-filled planet? We have tentative answers to some of these questions, and just wild guesses to others, but it is clear that exploring this interval of extreme, harsh stability in Earth's climate is fundamental to gaining a greater understanding of what happens when a system is pushed to its planetary boundaries, and how life persists, and thrives, nonetheless.

Setting the stage for a snowball

As discussed in Chapter 3, one of the first major snowball intervals that the Earth experienced, the Huronian Glaciation, was in response to the Great Oxidation Event. That one, occurring approximately 2.4—2.1 billion years ago, was driven by the evolution of oxygenic algae and the eventual saturation of the "rust machine" that was the early reducing Earth. Once the atmosphere started getting whiffs of free oxygen, a number of things changed—most notably for climate, the oxygen reacted with the prevalent methane gas in the atmosphere and oxidized it to carbon dioxide. Given how much more powerful a greenhouse gas methane is compared to carbon dioxide, the removal of this gas quickly cooled the planet, and geologic evidence supports a cool-down so extreme that the planet experienced extreme glaciation until a new carbon balance was set.

So why is the Snowball Earth interval from about 700 to 600 million years ago so notable, when we had already seen at least one before that? First, the Huronian glaciation was not likely fully global in nature and might have been more like an extreme version of recent Ice Age intervals that the Earth currently experiences. Second, it was driven by a biogeochemical event, the evolution, and proliferation of oxygen-producing organisms, and thus at least has a mechanistic hypothesis as to its origin. Finally, the Sturtian glaciation was followed by an interval that many paleoclimatologists call the "Boring Billion." This was an extremely long interval of extreme planetary stabilization from about 1.7 to 0.7 billion years ago. Based on geochemical, isotopic, and paleontological data, it seems as if absolutely nothing happened during this time. No major changes in the ocean or atmospheric chemistry. No major evolutionary steps. No major climatic swings. It was as if the Earth, and all life on it, stood still for a billion years, waiting for something big to happen. But perhaps, lurking in the watery depths of the ocean, a long series of experiments in evolution was taking place. Experiments can tell us a lot about the evolution of life on this planet, and perhaps on other planets in the Universe.

The Boring Billion and the snowball set-up

Living as we do on a planet that is constantly changing, on a human lifespan (rotary dial phones?!?!) and an easily observable historical one, it is simply beyond imagining nothing much happening on the planet for a billion years. The rotary dial phone largely went extinct by the 1980s (although the anachronism of "dialing a number" persists), about 40 years before this chapter is being written. Kids can sometimes identify what this device is but hilariously fail at knowing how to use it (https://www.youtube.com/watch?v=oHNEzndgiFI). A billion years is 2,500,000 longer than it has been since the rotary phone went extinct. Even in geologic perspectives, this duration of apparent stability boggles the mind—the last billion years alone have seen the evolution of animals, the birth of forests and the colonization of continents, the rise and fall of dinosaurs, and the evolutionary arc that eventually led to the evolution, and extinction, of that rotary dial phone. But recent research has indicated that perhaps a far more complicated evolutionary dance was occurring in the depths of the ocean, one which would end up shaping the modern planet as we know it—evolutionary experiments that resulted in the first animals to inhabit the planet.

Plant life during the Boring Billion

Prokaryotes and eukaryotes were both happily existing in the oceans from 1.8 to 0.8 billion years ago, albeit in an ocean that had less dissolved oxygen in it and thus likely more chemicals that are a challenge to modern organisms. These chemicals include more hydrogen sulfides and more dissolved heavy metals, both the result of a less oxidizing ocean. One additional difference is that marine life might have been forced to live a bit deeper in the water column than in the modern ocean because the UV-protecting ozone layer had not yet fully matured. The ocean itself does an effective job of absorbing UV radiation, so it is likely that the oceans would have been safe for organisms below about 10 m in water depth. Along with these differences, and the absence of abundant/any marine animals, there was little in the way of larger marine plants, such as the kelp forests and coralline algae beds that currently proliferate in the shallow subsurface of oceans around the world. It was an ocean of the minuscule, ruled by microscopic algae and bacteria that were photosynthesizing and chemo-synthesizing in waters that would be closer to those of the modern "Dead Zone" in the Gulf of Mexico than the vibrant and oxygen-rich water off Hawaii.

Even with these lower oxygen and higher sulfide waters, however, one would expect to see some evolutionary movement afoot. Particularly in the case of eukaryotes, which had already evolved 1.8 billion years ago to effectively photosynthesize and manage relatively complex biochemical reactions. But in fact, many studies are beginning to reveal distinct chemical signals locked in the rock record. These suggest that major evolutionary changes were occurring, but just weren't obvious because these early organisms left no fossil traces of themselves, given that they lacked shells, or bones. Evolution typically comes from environmental stresses, and new research indicates that this stress came not from the relatively low oxygen levels at the time, as organisms had already been juggling that issue for some time, but from the lack of other essential nutrients that come along with these lower oxygen levels.

Photosynthesizing organisms need many nutrients to thrive, including the major ones, such as carbon, phosphorus, and nitrogen, but also myriad minor ones that are typically in adequate supply. But the generally lower oxygen concentrations in the environment at this time seem to have limited the mobility or availability of a few of these minor nutrients. For example, careful geochemical analysis of pyrite (iron sulfide) minerals in shale rocks (https://www.utas.edu.au/research-admin/research-news/new-research-finds-boring-billion-did-promote-evolution-of-life) that formed at this time point to a shortage of selenium, copper, cobalt, molybdenum, zinc, and cadmium, all of which are necessary at some level at least for normal functioning (look at your bottle of multivitamins if you want confirmation!). These shortages likely sparked the process of endosymbiosis, where single cells ingested by other cells ultimately lead to more complex cells. This process is fraught with issues, given that it more likely leads to cell death than evolution, but if you had a billion years to keep at a task, it is bound to work at some point.

Not only was oxygen not static during this entire billion years but also that there is some rare fossil evidence of organisms that might look like marine plants to us—macroscopic seaweed-like algal clusters (https://www.nature.

com/articles/ncomms11500) whose appearance seems to coincide with these bumps in oxygen levels. Based on evolutionary modeling, at least it is likely that the major split between plants, fungi, and animals occurred sometime in the Boring Billion (https://bmcecolevol.biomedcentral.com/articles/10.1186/1471-2148-4-2), driven by the recurring experimentation with endosymbiosis. It may be that the subtle variations in oxygen levels and in the availability of micronutrients, combined with time, were critical variables that set the biological stage for events that occurred during the Snowball intervals and the subsequent Cambrian explosion of animal life on the planet.

It is interesting to consider Mars in this context. Mars might have experienced its own Boring Billion, or Boring Half-billion, having much the same geochemical starting ingredients as Earth. If so, and if prokaryotic life started there as well (or we swapped life via collisions), how far along the evolutionary road might have it gotten before its major freeze-over? And unlike the earth, Mars' deep freeze was for good. As we are clear evidence of, organisms on Earth made it through its Snowball phases, and indeed might have been tempered by the extreme conditions to be ready for evolutionary explosions when it ended. But on Mars, could organisms have persisted, lurking in the warmer, wetter, saltier subsurface of the planet, for the subsequent several billion years? Or might we only find their fossil remains, suggestions of life's Martian vibrancy indicated by isotopic shifts in remaining rocks records? Clearly questions for another day, but on Earth at least the stage is now set to test the limits of Earth's climate and life cycles through a very long glacial night.

The snowball concept emerges

The slow awakening of earth scientists to the concept of a Snowball Earth began with the discoveries of the Australian geologist Sir Douglas Mawson in the early part of the 20th century. Finding vast glacial deposits in southern Australia, he rightly contended that the continent experienced significant glaciation around 700 million years ago, but got the location wrong. At this time, the concepts of plate tectonics and continental drift were not established, so he assumed that Australia was at its same latitude as it is today and that seeing glacial deposits here implied global ice cover. He was right about the global glaciation part but wrong about the "convincing evidence" from temperate-zone Australia—the continent was actually located at more of a polar position at that time, where glaciation would not be a challenge to find based on the modern analog of Antarctica.

Regardless, these geological observations of widespread glaciation mounted, followed by a host of modeling and more sophisticated paleomagnetic studies. The modeling studies proved challenging—in the 1960s the great Soviet climate modeler Mikhail Budyko was able to simulate global glaciation but only if ice cover advanced to a low enough latitude—a runaway train of ice build-up, increased albedo and reflection causing cooling, followed by more ice, etc. (Fig. 4.3).

The problem with Budyko's model simulation however was that there was no way to turn it off. The Earth effectively froze over from pole to pole and at that time there was no known mechanism to thaw the planet again. Then along came the full development of plate tectonics, and with it, a way out of the icy conundrum—volcanic eruptions. The long carbon cycle eventually works to achieve an equilibrium, and in this case, the carbon dioxide released from sub-ice eruptions likely accumulated in the ocean and leaked into the overlying atmosphere, either via the rare open areas of the ice-free ocean or through cracks in the overlying ice cover. Once enough carbon dioxide accumulated in the atmosphere, its warming effect eventually beat back the ice cover, returning the planet to its thawed-out "normal" state. But this simple "global glaciation until volcanoes come to save the day" scenario glosses over many of the critical basic questions of timing, mechanisms, and controversies that revolve around this interesting period in Earth's history.

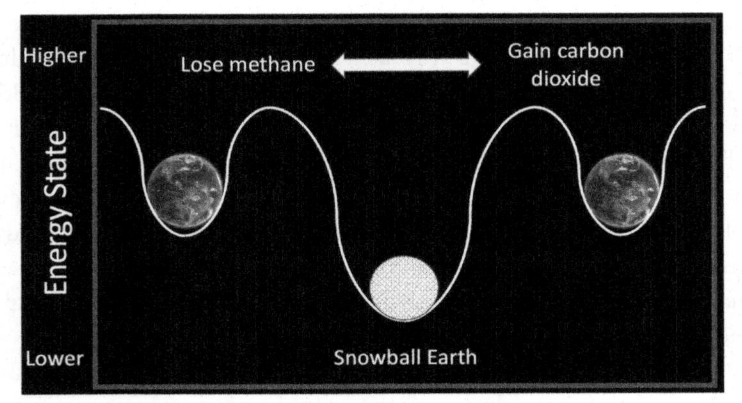

Figure 4.3 ***The Snowball Earth Climatic Well.*** The climatic inertia of a Snowball Earth scenario is difficult to break out of.

The first, long snowball

About 750 million years ago, after a billion-year interval of geologic boredom and tectonic stability, came perhaps one of the key events that triggered the first, 58 million year-long Snowball interval—the beginning of the break-up of the Rodinian supercontinent. What does a break-up mean, and how does it impact global climate? First, it is important to separate the two types of hard outer cover that the Earth has—oceanic lithosphere and continental lithosphere. These are often separated by a zone of steep contrasts, and in a simple way can be thought of as the "plates" that clad the Earth. Continental plates are usually very old, thick, and buoyant, and many have been around for billions of years, being pushed to and fro by the actions of plate tectonics like ships on a sea. In contrast, oceanic plates are very young by geologic standards (the oldest lithosphere is only 170 million years old), quite thin, very dense, and are heavily recycled by plate tectonics. For example, one large chunk of the continental lithosphere is the North American Plate, which spans from the polar region of northern Canada and Alaska down to the narrow Isthmus of Panama. This massive plate is mostly billions of years old, except for much of the western part that has experienced significant addition of material effectively scraped off the massive Pacific oceanic plate to the west. The boundary between the North American Plate and the Pacific Plate is extremely geologically active, marked by frequent earthquakes and volcanoes that are characteristic of this type of active margin. In contrast, the eastern edge of much of the North American Plate is marked by a smooth, geologically quiet transition to the Atlantic oceanic plate.

Rodinia is an example of a time in earth's history when much of the planet's continental lithosphere was stacked up in one single large landmass. The Rodinian Supercontinent was stable throughout much of the Boring Billion, perhaps aided by extremely weak plate tectonic activity at this time. But two things resulted from the restart of more active plate tectonics—the break-up of Rodinia into continental chunks that somewhat resemble the continents that we see today scattered more evenly across the globe and significant changes in the planet's carbon cycle. The break-up of Rodinia was initiated by rifting processes, where the continental lithosphere is heated, rises, and is literally torn asunder from beneath, with the gaps soon to be filled with brand new oceanic lithosphere. A modern analog is east Africa, whose string of great lakes mark the central part of a continental rift, and the tall mountains on either side of the lakes represent uplifted continental lithosphere. One of these tall mountains in Kenya, Mt. Kilimanjaro, is so high that it sustains a peak glacier, even though it is nearly on the equator! Sadly, this glacier is rapidly melting due to climate change, but nevertheless, it provides a snapshot of what happened when Rodinian uplift occurred. Imagine not just one Kilimanjaro but thousands, accumulating ice over the winter that never melts during the summer, and slowly but inexorably coalescing to form an albedo-fueled super ice sheet that marks the birth of the Snowball Earth.

While mountains were rising and accumulating ice at their tops, all of the new, low-lying oceanic lithospheres that were filling in the gaps left from rifting were still unusually hot, being freshly erupted, and quite buoyant because being relatively light. These new shallow ocean basins couldn't hold much water, which meant that the oceans lapped up even onto the low elevation areas of continental crust. These became prime areas of high evaporation, causing significantly more rainfall on the tops of the new mountains, further fueling ice build-up. Additionally, this extra rain caused an increase in the chemical weathering of these rising mountain chains, and because the chemical weathering of continental rocks extracts carbon dioxide from the atmosphere and sweeps it into the ocean, greenhouse warming was reduced at this time.

Geological rifting kicked in again, with new mountains growing glaciers, shallow seas causing bigger glaciers, and chemical weathering absorbing carbon dioxide from the atmosphere simply too strong of forces for the normal medium-term carbon cycle to counteract. Because rifting of Rodinia likely took place over tens of millions of years, the other usual feedbacks in the Earth's carbon system couldn't keep pace with the ever-building high albedo ice cover that was creeping from pole to equator. The albedo effect is particularly important to consider here. Ocean water absorbs roughly 88% of the solar energy that hits it, with only 12% reflected back out to space. Land absorbs a bit less, something like 60%—90% of incoming energy depending on the mineral types that comprise rock. Those absorption numbers are likely elevated because they included vegetated landscapes, and the continents had little to no vegetation at this time because the ozone layer had not yet been strengthened to protect organisms from ultraviolet radiation damage. But compared to ice, these absorption numbers are huge—depending on the thickness and whiteness of ice, it absorbs only 20%—45% of the sunlight hitting it, and the remaining 55%—80% is reflected right back to space. So, as the ocean and land surface were slowly but surely covered in ice (Fig. 4.4), the global reflectivity went up and the

Figure 4.4 ***Ice-Covered Landscape.*** Ice grew from mountains all the way down to the oceans in the lead-up to Snowball Earth conditions.

global surface temperatures went down. And with a Sun that still hadn't fully powered up by then (being roughly 6% dimmer than today), there was little to stand in the way of the freeze. Soon, a near-global ice sheet covered the planet, and wouldn't lose its grip for another 58 million years.

Snowball, slushball, and other remaining questions

Before diving into the unique submarine biological revolution that may have been taking place while the planet was in the grips of a freeze-over, it is important to consider whether, in fact, Earth was truly a Snowball or something else. A number of counterarguments have been posed to the Snowball hypothesis, from simple disbelief when it was first proposed to more subtle arguments about the thickness or nature of the ice cover. The disbelief came from a simple question that most geologists are trained to start with—what, or where, is the modern analog? We are trained to look for modern equivalents of rocks or processes to describe conditions of the past Earth stored in the geologic record. So, when examining layered sandstone rocks in Zion and Bryce Canyons in the Southwestern US, geologists first have to decide whether they were formed underwater or on land. In the case of these deposits, they have clear patterns of ripples or waveforms, but they are impossibly thick to likely be formed underwater due to the forces required. Thus, the hypothesis is that these deposits are the remains of a massive dune field that existed in this area at the time, similar to but likely larger than the Sahara. Ripples on the surface of Mars spied by the Curiosity Rover, however, include multiple clay clasts that on Earth only form when water is present, and thus the interpretation is that these ripples were formed by the ancient, now evaporated, lake that formed in the basin of the Gale Crater on Mars. In this way, we are depending on the physics of rock-forming processes being relatively similar no matter where you are—even on Mars!

But a planet covered completely by ice when it had none before? And with an even brighter Sun than during the first Snowball, which occurred when methane was stripped out of the atmosphere upon the first whiff of oxygen after the Great Oxidation Event? Might there be a more plausible scenario that could still explain glacial deposits near the equation while maintaining a general condition that was more, well, realistic? Here is where the Slushball Earth hypothesis comes in. What if the planet had significant glacial ice on land and some surfaced sea ice but was not completely iced over?

In this scenario, much of the ocean would retain its normal function of evaporation and condensation, which is consistent with observations of a still-active hydrological cycle during this time. A fully glacial planet would have a cold and dry atmosphere. And again, with the analogs, this is why our Ice Age climate causes the significant expansion of deserts on land, because of a dry atmosphere. But during the Snowball interval, there is plenty of evidence for water transport of sediments as well as glacial transport, and a halfway (Slushball) environment would be more consistent with that evidence. Also, there were no recorded extinctions at this time, which might indicate an ocean that functioned not completely different than it had before the supposed ice-over. Then again, there really wasn't much to go extinct anyway at this time, so this Slushball evidence is a bit less definitive. Nevertheless, a Slushball does provide a compromise scenario that many earth scientists felt more comfortable with when the concept first emerged.

Ultimately, however, newer evidence of timing and climate dynamics now favors a full Snowball scenario. A Slushball planet would have allowed carbon dioxide produced from submarine volcanoes to leak back out to the atmosphere more readily, thus resulting in any glaciation being short-lived, cut short by the normal stabilizing feedback mechanisms of the carbon cycle. But the Snowball conditions were anything but short—as noted earlier, the first one lasted an incredible 58 million years. This glaciation is also known as the Sturtian glaciation, named from the location, the Sturt River Gorge in South Australia, where the first evidence for it emerged. And on top of that, when it finally ended, it did so abruptly, and the earth experienced a ~10-million-year hot spell before the next, shorter (5–10 million years—the Marinoan glaciation) Snowball kicked in. This type of climate dynamic is similar to what we, again, observe in the recent Earth analog of the Ice Age cycles, where glacial conditions abruptly end due to rapid injections of carbon dioxide into the atmosphere following the orbital changes that drove the planet into Ice Age conditions in the first place.

Questions about dynamics during the Snowball Earth interval still exist, many provoked by the newer timing estimates. Why was the first Snowball 58 million years long, and the second one so much shorter? Why was there a strange 10-million-year warm interval between the two? Why weren't there more? Even in the coldest of years during Ice Age intervals the planet overall is much warmer than Snowball Earth and doesn't ever experience the scale of positive feedback necessary to promote the equator-ward march of ice sheets seen back then. Clearly, the geologic chapter is not yet closed on the driving mechanism(s) of the Snowball Earth, but one thing we know for certain—soon after the end of the second one, the Earth experienced perhaps its greatest evolutionary diversification episode ever, driven by the strange submarine bedfellows that an iced-over ocean created.

Evolutionary action after the thaw

Most studies of the beginning of the evolutionary explosion point to an interval shortly after the melting of the first Snowball glaciers (https://cosmosmagazine.com/earth/earth-sciences/how-snowball-earth-gave-rise-to-complex-life/). This is the period that saw the global rise of algae, around 659-645 million years ago. Research led by scientists from the Australian National University (https://www.nature.com/articles/nature23457) found molecular fossils indicating a significant diversification and expansion of algal species complexity at this time. These molecular fossils are actually organic molecules that are unique indicators of biological complexity—the sterols that the scientists found in these carefully dated rock layers are those that only multicellular organisms (eukaryotes) can make. This marine plankton proliferated in the post-Snowball oceans, perhaps setting the stage for the more complicated evolutionary tricks to come.

Many possibilities exist for why it was only at that time that eukaryotes significantly expanded. The most straightforward explanation has to do with the remarkable changes in global nutrient dynamics that emerged after the massive glacial melt. Critical biological nutrients like phosphorus were in such a high supply at this time (https://www.nature.com/articles/nature20772) from rapid and extreme weathering of the finely ground rock left behind after glacial retreat on land that organisms could easily experiment with many different evolutionary variations. High nutrients could also tip the ecosystem scales from a state of biological competition to one of cooperation, providing a type of environment that then favors biological complexity over pure efficiency. Perhaps various precursor steps to the eukaryotic expansion also occurred when the oceans were iced over, including the conditions of ocean stratification (layering) that ice cover promotes, which created stable yet closely spaced environmental zonation that encouraged biological innovation at these submarine interfaces.

With the "Rise of Algae" came more complicated food webs, which are more efficient at the transfer of nutrients and energy through ecosystems, which then supported more complex, and larger, organisms. Indeed, at this time the rock record reflects the first biomarkers for animals—sponges and predatory rhizarians (https://www.nature.com/articles/nature07673). Thus, sometime during the interlude between the two global glaciations and the end of the second and final one (the Marinoan), animal life emerged on this planet for the first time. This huge evolutionary leap also indicates that oxygen levels were high enough in at least some ocean systems to support animal evolution, which is no small feat given the extremely high oxygen demands of animals. Recall that, unlike plants, all animals burn consumed sugars in the presence of oxygen to produce energy, and thus our very origination and ultimate global expansion as a Kingdom depends on this one gas.

Early biota caused the first free oxygen on the planet, a gas that never left after that point, and ushered in an ice age. A massive and prolonged ice age later forged the environmental conditions necessary for the ultimate evolution of the true oxygen consumer—animals. And as we learn later the multiple recent Ice Ages helped shape the arc of hominid evolution, which has resulted in this particular hominin to be writing this book about climate change and life —ice is a recurrent character in this story of life on Earth! Before we unfold the rapid evolution and proliferation of life in the seas and on the continents of our planet in the next chapter, we will wrap this one up with a quick diversion, or dive, under the oceans and ice in search of life—but these oceans are on planets circling other stars and this ice is on a large ice-covered moon 628 million miles away.

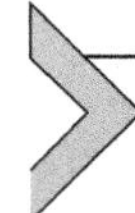

Snowballs, water worlds, and the search for life "on the edge"

Sometimes, to understand the complex relationship between climate, life, and evolution, we have to go to strange places. Those scientists who work to unravel the origins of ancient chemical-consuming life forms will go to submarine volcanoes at the ocean's depths, UV-seared deserts of the high Andes, or deep caves in Europe and the US. These environments somewhat resemble the conditions that presided on the Earth billions of years ago, before the Great Oxidation Event and the onset of an oxygen-dominated world. But to understand the unique conditions that presided on our Earth during its icy Snowball periods, or those that might have existed in our early warm history when we were closer to the Water World of Kevin Costner fame (or infamy?) than now, we have to go far afield. Indeed, this journey for life's origins in strange places will take us to lakes frozen under the Antarctic ice sheet, the icy moons of Jupiter and Saturn, and the many planets in our galaxy that are literal Water Worlds, covered by a planet-spanning ocean. By exploring these extreme environments on this and other planets, we gain a greater sense of the flexibility, creativity, and wonder of life on our own planet.

Water worlds

OK—full disclaimer to start. The premise of the 1995 postapocalyptic climate change (the so-called "CliFi" genre) film "Waterworld," starring Kevin Costner, along with its 2004 CliFi film cousin "The Day After Tomorrow" starring the Kevin Costner look-alike Dennis Quaid, is ludicrous. In the former, climate change melts all continental ice, including the ice sheet on Antarctica, and the planet becomes an ocean-covered globe with nearly no land to be found. In the latter, a hiccup in the climate system driven by human-induced climate change ends up covering much of the globe with ice instead. Kudos to the writers and directors for grappling with climate change in these films, and for using a bare seed of scientific fact—if unchecked, climate change will indeed melt all continental ice, and there are indications that as the planet warms, counterintuitively the northern Atlantic region might get colder—to drive both of these Hollywood Blockbuster wanna-be's.

Both films were widely panned, because, well, they were awful, and I would like to think that the audiences knew that (1) even if all continental ice was melted, the sea level would only rise by tens of meters, eating away at only a small portion of a continental area, and (2) the temporary, and local nature of North Atlantic cooling would not spark 100 m thick ice sheets to terrorize New York City like some wrathful icy Godzilla. Yet both films in a strange way related to very real conditions on Earth during the past, and those that are increasingly likely scenarios on large moons circling other planets or planets circling other stars. If we are to understand the full range of planetary conditions that can lead to the formation and evolution of life, we need to move beyond the silver screen and seek out models and examples that can explain our real world.

Nutrient cycling on a water world

On Earth, there have always been several nonnegotiable ingredients for the support of life, and the slow evolutionary pathway from Archaea to Homo Sapiens. Water is obviously one of them, but water has been plentiful for much of Earth's history. But one essential ingredient that has always been in short supply, and whose geochemical cycle results in it frequently being sequestered in forms that are not readily available to biota, is phosphorus. The Earth never started off with much of this element and as long as the planet has been oxygenated, much of the phosphorus made available by continental weathering is readily trapped by iron oxidation (e.g., Filippelli, 2016). In fact, the only new source of phosphorus to the global biosphere is via the chemical weathering of terrestrial minerals. So, if Earth were a "water world" planet, covered completely by the ocean with no possibility of terrestrial supply of phosphorus, would life have ever proliferated here? This is not just a rhetorical question, but one that is relevant to conjecturing about life on early Earth and on other planets, including those that have been detected that seems largely to be covered with water.

In a combination of geochemical chamber experiment/thought piece on exobiology, Syverson et al. (2021) ask this question and come to the intriguing conclusion that a water world would indeed have enough phosphorus released from an unlikely source to sustain life—submarine basalt weathering. Many models of astrobiology are based on Earth analogs, and certainly, reasonable arguments could be made that the lack of terrestrial phosphorus weathering on a water world would significantly reduce phosphorus weathering production, perhaps making it as much as 1000-fold lower (Glaser et al., 2020). In the deep ocean on Earth, hydrothermal vents and the oxides that form from emitted vent fluids are certainly net consumers of phosphorus, and massive ones, comprising approximately 20%

of the marine export flux for phosphorus. But as Syverson et al. (2021) argue, if the water circulating through submarine vent systems and the deep ocean into which they emit was anoxic, the oxidative phosphorus loss factor be removed. Additionally, based on lab experiments, the weathering of submarine basalts in these anoxic conditions might produce significant amounts of phosphorus (Fig. 4.5).

The global phosphorus cycle, or more accurately that before humans began perturbing it with the widespread application of chemical phosphate fertilizers, is simple. Having no gas phase like the nitrogen cycle, the only new source of phosphorus to global biogeochemical cycles is through the slow chemical weathering of continental rocks. Once released from its mineral form, phosphorus is transformed in the soil profile into a variety of fractions that have been chemically defined and crudely reflect their reactivity—"occluded" phosphate bound to oxyhydroxides, organic phosphate, and "non-occluded" or adsorbed phosphate (Filippelli, 2016). Natural soil ecosystems are highly efficient in retaining phosphate as it cycles between its various geochemical forms. But the significant loss of phosphate from the landscape occurs with land disturbance, such as glaciation and deglaciation, and in modern chemical agricultural systems where fertilizers are typically overapplied, and the phosphate runoff causes significant water quality issues through eutrophication.

A portion of the phosphorus that eventually makes its way to the ocean is unreactive in the marine environment, but that portion that is reactive drives marine biological productivity on longer timescales, and thus modulates and net organic carbon export from the ocean. This is not to say that the internal marine phosphorus cycle is simple, and indeed, there is significant action around suboxic and anoxic portions of the ocean within the water column and in sediments. Long ago termed the "iron-phosphate pump," oxide-bound phosphorus is released in sub/anoxic systems and becomes dissolved phosphorus, available for biological uptake (Wheat et al., 2003). Thus, a system that is sub/anoxic can lead to significant internal recycling of phosphorus and to higher net carbon export should other limiting nutrients be present in adequate concentrations. This is the scenario that has been proposed for driving significant organic carbon burial at various times in the past, with the net result being a loss of atmospheric carbon dioxide and lower global temperatures. But until now, little attention has been paid to what might happen if deep ocean water itself was anoxic (perhaps because this is not a situation seen today?), and circulates through seafloor hydrothermal

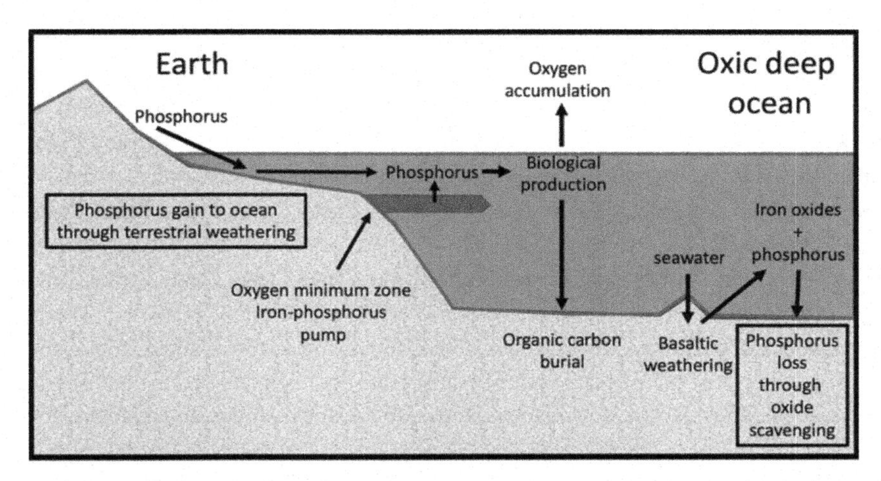

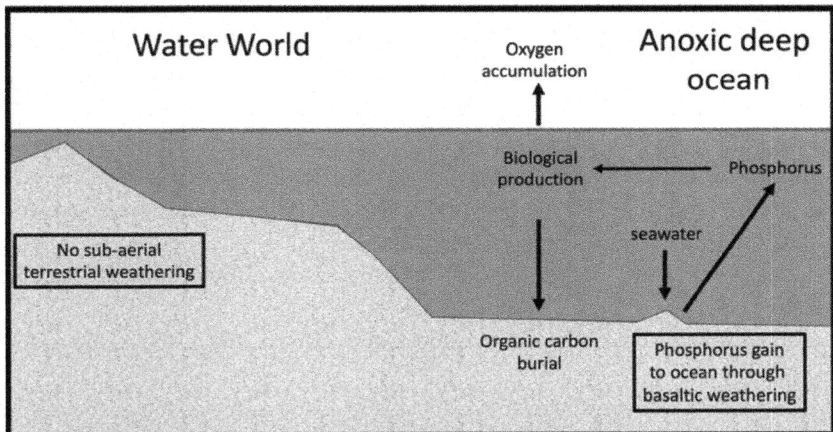

Figure 4.5 ***Phosphorus Cycling on a Water World.*** A schematic of the global phosphorus on today's Earth (A) and a "Water World" planet (B). The key difference between the two scenarios is that the vast majority of new phosphorus to the marine system fuels productivity comes from terrestrial weathering of exposed landscapes on today's Earth (A), whereas phosphorus from a "Water World" might be provided by submarine weathering of seafloor basalt under conditions of deep ocean anoxia in a hydrothermal spreading system (B).

systems as a weathering agent for phosphorus, instead of an oxide-rich phosphate scavenging environment as it is now.

To determine the extent and rate of anoxic phosphorus weathering from seafloor basalts, Syverson et al. (2021) conducted a range of long-duration chamber experiments. The experiments lasted approximately 2 months, and the results were striking—indeed, not only is the phosphorus weathering release high in anoxic conditions, but it is comparable to the current weathering release of phosphorus from continental rocks. The reaction of dissolved carbon dioxide being consumed during submarine basalt weathering is balanced by the weathering release of phosphorus and the consequent uptake and eventual burial of organic carbon. One product of this reaction is oxygen. In these conditions, this process may yield as much biospheric oxygen as that produced on modern Earth, and significantly more than the oxygen production soon after the Great Oxidation Event (Ozaki et al., 2019).

Interestingly, although a number of studies have focused on phosphorus cycling in the early anoxic ocean (both pre-Great Oxidation Event and at various intervals of the Paleozoic), they've been using the model of the modern anoxic zones as a frame. In modern anoxic systems, significant phosphorus recycling occurs from oxide-bound phosphorus as it is dissolved in the water column or sediments. This process drives the assumption that in a mostly anoxic ocean, the phosphorus concentration would be higher because of this process alone, and yet the origin of that marine phosphorus is assumed to be terrestrial weathering. Submarine basalt weathering has not typically been included in calculations of the phosphorus cycle at these times because of the assumption terrigenous weathering dominates the global phosphorus cycle.

The second implication of this study is that an exoplanet with significantly more water (per mass) than Earth may not only have adequate phosphorus for life, but also a carbonate-silicate cycle with feedbacks not altogether unlike those with exposed landmasses. This becomes a critical and perhaps groundbreaking finding for astrobiology, where the habitable zone is defined as the distance envelope around a star where a black body planet would have a temperature between 0 and 100°C—i.e., a planet with liquid water. The distance window is defined by the energy output of a given star, and would in reality be modified by the greenhouse gas composition of a planetary atmosphere and the albedo of that planet. But the classical concept of a habitable zone has been considered moot for planets with global oceans (Abbot et al., 2012; Foley, 2015), because it was presumed that the normal atmospheric stabilization of carbon dioxide in the atmosphere was achieved by carbonate-silicate weathering cycle (i.e., the Kasting et al., 1993 model), and a planet without exposed landmasses for silicate weathering would not qualify for a robust self-regulating climatic system required for the slow timescales of biological macroevolution. This ruled out a tremendous number of candidate exoplanets, as "water worlds" are common (e.g., Mulders et al., 2015). It seems that there is some justification to rule them back into the realm of potential life-harboring planets.

Icy moons and Antarctica

Two moons in the Solar System are covered completely with water ice, floating on liquid oceans. We know this because of various space probe fly-bys of these lunar bodies. One, the large moon Europa (Fig. 4.6) orbiting the monster planet Jupiter is easily observable from Earth with a pair of binoculars. Find Jupiter on your astronomy app (it is the brightest "star" in the night sky after another planet, Venus), look up with your binocs or telescope, and see four bright dots arrayed in a line on Jupiter's orbital plane—the second one away from the planet is Europa. The other, Saturn's moon Enceladus (Fig. 4.7), is not easily observable with a backyard telescope, but is similarly covered in a smooth layer of ice—indeed, these two moons are considered the two smoothest solid objects in the Solar System. Europa is

Figure 4.6 ***Europa, One of the Large Moons of Jupiter.*** Europa is covered by a layer of ice, but has a deep liquid ocean underneath and is one of the prime candidates for the exploration of life beyond Earth.

Figure 4.7 ***Enceladus.*** Enceladus is one of Saturn's moons, covered by a layer of water ice but with a liquid ocean beneath.

large for a moon, being about ¼ of the radius of the Earth, and Enceladus is small, being only about 15% of the radius of the Moon.

Both orbit their planets very quickly, making a full orbit in a few days or less, and both are "tidally-locked," meaning that like our own Moon they do not spin on their own and keep their same face toward their planet at all times. The monstrously huge planets that each of these moons orbits have such a strong gravitational pull that the tidal potential is quite high, meaning that the forces acting on the moon's surfaces is enough to cause significant fractures and rearrangements of their surface ice cover.

Both moons appear to have some degree of tectonic activity, either from the moon's internal heat sources alone or in combination with the tidal forces exerted on them by their huge planetary neighbors. They also contain a tremendous amount of water. Because they are so far from the sun, and the planets that they orbit emit radiation but not nearly the power of the Sun, their surfaces are frozen. But the tectonic and tidal forces acting on these moons causes internal heating that is adequate to sustain a significantly deep liquid ocean under their icy surfaces. In the case of Europa, the ocean (including the icy crustal cover) is estimated to be about 100 km deep, much deeper than Earth's average depth of 3.8 km. Both of their icy shells are dynamic and crack and open, allowing mineral-rich ocean water from below to emerge on the surface in the case of Europa and to explode occasionally over the polar area in the form of "cryogeysers" in the case of Enceladus.

For decades now the intriguing idea has been circulating among the astrobiology community that Europa in particular would make an excellent target for exploring whether conditions in the subice oceans are amenable to life, or indeed if life has already evolved there. Certainly, the same mineral-rich, warm, liquidy conditions occur in the Europa depths as did on Earth and Mars billions of years ago. Furthermore, the icy surface layer is more than enough to protect organisms from the vicious radiation that spews out of Jupiter—strong enough to cook electronics of probes, and thus easily toxic to exposed life as we know it at least. For unprotected humans, the radiation dose received at Europa's surface in a single day would cause severe illness or death. There is substantial debate about the thickness of the outer ice layer, with one camp leaning toward a thick ice model of 10—30 km thick and another the thin ice model of about 200 m thick. Even at the thin end, radiation protection would be complete for any organism swimming or floating in the subsurface ocean.

Which of the two models is correct, thick ice or thin ice, has huge ramifications for the approach to exploring life in Europa's depths. Either would be a daunting challenge, of course, especially given proven challenges that Mars surface explorers have faced running without real-time control. In the case of Mars, it takes 10—40 min to get a message from Mars and then transmit instructions to be acted upon back to Mars. The difference in time is due to the orbital configurations of the two planets relative to each other as they circle the Sun (anyone who has read or seen "The Martian" knows how awkward this time delay can be). For Europa, that back and forth time is about 80 min or longer depending on the orbital configuration. It is difficult enough to get exploration progress accomplished on Mars with unmanned surface robots and requires significant intelligence to be built into these robots, and obviously becomes more challenging for an unmanned Europa mission.

For a "thick ice" Europa, the search for evidence of life under the ice would need to focus on either the fracture areas that riddle Europa's icy crust or areas with smoother surfaces, which would indicate a lack of asteroid bombardment and thus young ice. These areas could be explored by surface sampling and either in-situ analysis, similar to what the Curiosity rover performs now on the surface of Mars, or sample return, which is an objective of the newer Perseverance rover. For the latter approach, samples would be taken and sealed and stored in containers to be left on the surface and picked up and returned in a follow-up expedition to then be analyzed bringing the full analytical capabilities available in labs around the world. A surface rover with substantial in-situ analytical capabilities similar to Curiosity is literally decades away from construction, and launch, as it is only in its planning stage at the present (https://www.jpl.nasa.gov/missions/europa-lander). No substantial plans have been advanced for a sample return mission.

For a "thin-ice" Europa, the search for life in the underice ocean might be much more direct. Using technologies available currently, we are easily able to drill, or melt, through 200 m of ice in a day (https://www.nature.com/articles/526618a). In a very simplistic approach, one could imagine a landing craft (not a rover) that deploys a nuclear-tipped probe to quickly melt through the ice. One advantage of this approach is that it likely would also sterilize its pathway forward, ensuring that pesky microorganisms that might have survived the trip all the way from Earth to Europa might be killed so as to (1) avoid contaminating Europa and (2) avoid yielding some false positive where we get evidence of life only to find that it was a hitchhiker from Earth! Once through the ice, a tethered submarine-type craft could be deployed to visually observe the environment, capture sound waves, and perform biogeochemical analyses. All of these devices are readily available in any normal geochemical field lab at present, and thus need only be adapted for travel and deployment and shielded from radiation during travel and on the surface. In this way, a continuous, albeit delayed, the signal could be transmitted between Earth and Europa, with instructions coming from Earth and data from Europa. Of course, this all sounds easy, and the revelation of finding life in the subsurface ocean of Europa would be earth-shaking. But the coordination, cost, and luck involved in a successful mission will prove to be a tough combination to balance.

In the meantime, a slightly less ambitious but nevertheless exciting set of missions are in the planning or deployment stages. These include flyby remote observation missions by NASA, and by the European Space Agency, and an intriguing private venture. This latter is currently in the planning stages to "sniff" material erupting from Enceladus' cryoglaciers to assess signs of potential subice life. This is a privately funded effort (https://en.wikipedia.org/wiki/Breakthrough_Initiatives) that could be mobilized quickly and at a relatively low cost and would include onboard devices to measure the biochemistry of material that is emitted from Enceladus over a number of flybys, increasing the chance of capturing these moderately frequent eruptions in action. Additionally, the ice layer on Enceladus appears to be very thin (tens of meters https://blogs.agu.org/geospace/2016/06/21/saturn-moon-enceladus-ice-shell-thinner-expected/) and thus ice-penetrating radar could be employed to examine the physical structure of the subsurface ocean should a lander be deployed during a flyby. Much less can go wrong with probes examining lunar surfaces than those that have to autonomously land on a surface, without crashing!

Antarctica and the challenge of sterility

How does the discussion string from Snowball Earth to icy moons end up on Antarctica? It all comes down to technology, sterility, and exploring what happens if evolution is left to its own devices in a deep, dark, subice lake. Evolution is the process of individuals, and ecosystems, responding to external pressures in such a way as to enhance species survival and genetic transfer. As we have discussed throughout this book, those pressures are not only one-direction—consider the evolutionary leap that made oxygenic photosynthesis possible, and the resulting chain of environmental impacts, including oxidation of the Earth's surface during the Great Oxidation Events and the temporary climatic crisis following this, leading to the subsequent Snowball period. These push-pull dynamics between environment and evolution have functioned on Earth, and likely did as well on Mars, and Europa, should life have developed there as well, but they typically are only captured in the fossil record. What if you could see them in real-time? Or at least see their products in a real-life experiment that can be observed? That is where Antarctica, and its subice lakes, come into play.

Antarctica was not always a barren dry wasteland of thick ice sheets and happy penguins. It was only several tens of millions of years ago that the tall peaks of the Transantarctic Mountains reared up next to verdant forest-filled valleys and valley lakes teaming with life. Now, because of significant polar cooling that subsequently settled into Antarctica, those valleys and lakes are covered by kilometers of ice. But surprisingly, buried in certain deep valleys below all that ice, lie liquid lakes, sealed off from the rest of the global ecosystem evolution for millions of years. A literal subice Land of the Lost, where there is a potential that ancient species, sealed away for millions of years, have survived, and have even followed the typical pathway for ecosystems in their own isolated evolutionary pathways.

Lake Vostok, the largest of the hundreds of subglacial Antarctic lakes, was discovered in the 1970s via remote sensing. The lake itself has an average depth of meters, under 3 km of ice, and sits on a layered sedimentary deposit underneath. It was thought that the extreme pressure from that weight of ice alone causes the deepest portion to melt, but there are also some tectonic sources that provide subice heating. The ice cover in this part of Antarctica sealed the lake off about 15 million years ago, but the lake water itself is not likely that old—even the thick ice of Antarctica moves, and thus the sheet at the top of the lake likely freezes some of the liquid water at its base and moves away, leading to more new ice, which contributes water via pressure melting but also captures some of the lakes via freezing and carries it away. In this way, it is not likely that any of the original 15-million-year-old water remains, but nevertheless, the entire system is likely more environmentally isolated than any other ecosystem on Earth.

Attempts to explore the lake water itself, and unlock its potential biological mysteries, have been ongoing since the 1990s (Fig. 4.8). The justification for exploring these environments is pretty straightforward—in what ways might ecosystems evolve if largely isolated from the outside world for more than 15 million years, and what climatic history might remain in Lake Vostok's underlying sedimentary layers? The latter justification is important for paleoclimatological reconstructions of the Antarctic polar region, but it is really the first one that is critical to understanding how ecosystems evolve when isolated. In this way, the subice conditions of the Snowball

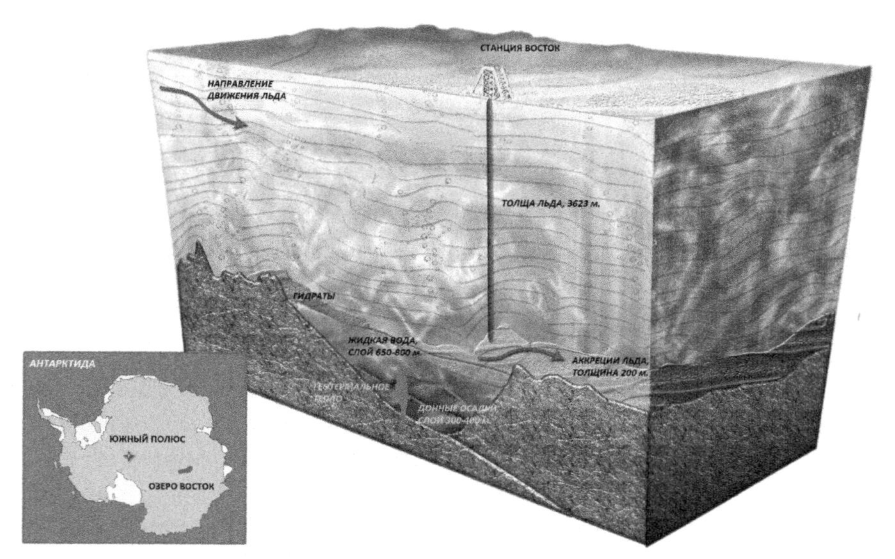

Figure 4.8 ***Lake Vostok and Scientific Drilling Targets.*** Lake Vostok deep under Antarctic ice has been the target of sub-ice exploration to assess life in this type of environment.

Earth, the subsurface aquifers of Mars, and the ocean depths of Europa all come together. And this then comes down to two exploration challenges. First, how do you drill down and place probes or extract samples from these isolated environments? Second, how do you do this while ensuring that your samples really do comprise organisms ONLY from that environment and that you haven't forever contaminated the environment for later unequivocal studies by inadvertently introducing exotic species from the exploration process itself into new, virgin, and potentially highly vulnerable ecosystems? In this way, direct exploration of the chemistry and biology of Lake Vostok is considered not only a key activity to get right the first time, but also a trial run for doing the same thing on Mars, Europa, or Enceladus.

As of 2022, the entire scientific effort of cleanly and safely extracting and analyzing a sample of Lake Vostok water can be characterized as exactly the opposite of what we should do when conducting these efforts off Earth. The efforts to drill down and then extract samples have been fraught with technical challenges, marked by international discord, subject to scientific debates, and met with environmental outrage. The drilling itself was conducted using kerosene and freon as antifreeze drilling lubricants to allow a drill to quickly pass through kilometers of ice. But these organic solvents are actually contaminants in pristine environments like this one and have resulted in a host of conflicting scientific findings. Additionally, the Russian efforts did purportedly breach Lake Vostok, but it is not at all clear whether the samples they recovered and analyzed are "clean" or even representative of lake water itself. Their efforts and those of other scientists have identified thousands of DNA indicators, including some related to larger organisms like fish that could be currently living in Lake Vostok. Or, as some argue, this is simply contamination from the drilling equipment, which was neither properly sterilized nor were the operations conducted in a way that a biologist would consider appropriate should they be doing these same experiments in a lab. The project has also been fraught with funding challenges, and the calls from various environmental groups to cease operations until sound technologies and approaches that maintain the pristine environmental conditions of Lake Vostok be developed and deployed. If Lake Vostok was a test of our ability to do these types of operations in a significantly more challenging off-world setting, we have clearly failed. But many lessons can be learned from this failure, and many scientists and engineers are tirelessly working to try to employ those lessons to help unlock the mysteries of the evolution of life and climate on our planet, and elsewhere.

Summary

Like another seeming "Game Over" moments in Earth's long and tumultuous climatic and evolutionary history, it is a testament to planetary resilience that the Earth recovered from its Snowball period. And not only just recovered, but was forever transformed as a unique evolutionary incubation was occurring under the ice that was soon to set the stage for the greatest evolutionary radiation period that has ever occurred—the Cambrian Explosion. Like the Great Oxidation Event and Boring Billion that proceeded it, and the multiple mass extinction events that the Earth has experienced since, the Snowball interval provides insight into how complex Earth systems function, and recover from shocks to achieve new states of balance. It is intriguing to consider that, gazing out at the night sky and all of the galaxies, and stars in those galaxies, and unseen planets circling those stars, you might be seeing some wrapped up in their own Snowballs, or Boring Billions, or Waterworld states. If observed by us, would we know that there might be life in them? Well, geologic and exobiological discoveries over the past several decades have made it increasingly clear that yes, these unlikely candidates very well might have life, and ecosystems, surviving, thriving, and evolving on their surfaces. Every new discovery, both down here and up there, seems to bring us more assurance that we are not likely alone, but equally, that this planet is a wonder of resilience, and we certainly occupy the best real estate around.

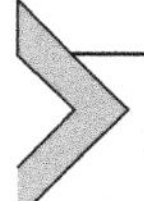

Did you know that?

That NASA has a position of "Planetary Protection Officer?" (https://sma.nasa.gov/sma-disciplines/planetary-protection)

This person heads up the Office of Planetary Protection at NASA, and their main objectives in terms of space exploration are to:

> *Carefully control forward contamination of other worlds by terrestrial organisms and organic materials carried by spacecraft in order to guarantee the integrity of the search and study of extraterrestrial life, if it exists.*
>
> *Rigorously preclude backward contamination of Earth by extraterrestrial life or bioactive molecules in returned samples from habitable worlds in order to prevent potentially harmful consequences for humans and the Earth's biosphere.*

That Europa Lander mission planning is well underway?

In 2020, over 400 scientists and engineers met to further solidify the mission and operational planning for a proposed lander mission to Europa. The overall project, called Europa 2050, calls for a multi-prong investigation for life above, on, and below the icy surface of Europa (Fig. 4.9). The first step of the program is to land a vehicle on Europa's surface (Fig. 4.10), and determine whether any signs of life can be detected in samples collected from 10 cm beneath the ice surface. This depth might be adequate for capturing signs of the complex chemistry of fluids that come from the ocean below, while also being deep enough to protect any potential biochemical indicators from the extreme and damaging radiation that is emitted from Jupiter.

The lander would include an onboard laboratory that is similar to the design of the robotic Martian Landing laboratories and rovers. An array of biomolecular detectors and spectrometers would be present for complex geochemical surveys of the ice materials. Additionally, it is possible that cameras, a microscope, and an onboard spectrometer would be available for visual analysis of material and for detecting any seismic activity within Europa.

The samples would be analyzed by a miniature laboratory within the robotic lander, similar to the way samples on Mars have been studied by landers and rovers on the Red Planet. In addition to its onboard chemical analysis lab, a Europa Lander mission might also carry a microscope and a camera, along with a seismometer to detect geologic activity such as eruptions or the shifting of Europa's ice crust.

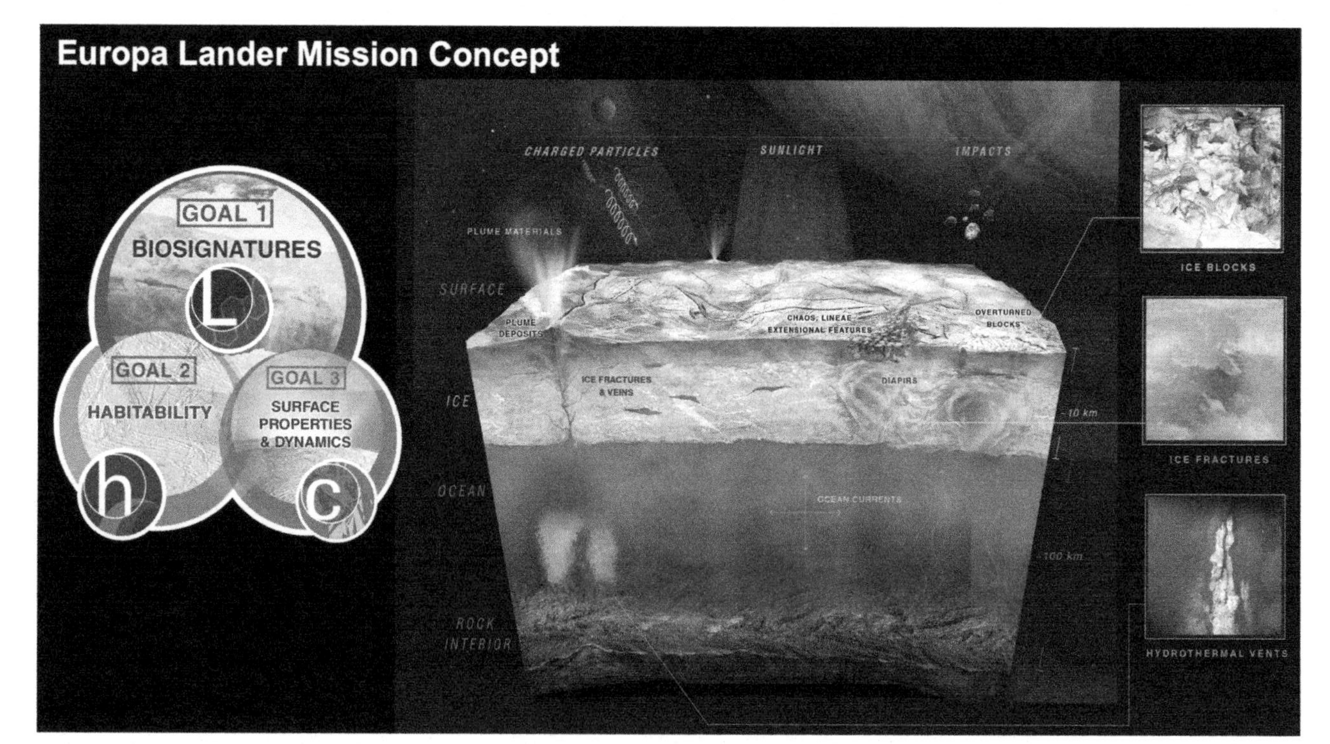

Figure 4.9 ***Expedition Plans for the Europa 2050 Mission.*** Scientific goals for a proposed mission to Europa to explore for life beyond Earth.

Figure 4.10 ***Proposed Europa Mission Landing Vehicle and Laboratory.*** The landing vehicle is designed to handle a smooth insertion even onto rough icy terrain, and from there to act using various robotic tools to detect a sign of life from the Europan ice.

References

Abbot DS, Cowan NB, Ciesla FJ. Indication of insensitivity of planetary weathering behavior and habitable zone to surface land fraction. The Astrophysical Journal 2012;756(2):178. https://doi.org/10.1088/0004-637X/756/2/178.

Filippelli GM. The global phosphorus cycle. In: Lal R, editor. Advances in soil science-soil phosphorus. CRC Press, Francis & Taylor; 2016, ISBN 9781482257847. p. 1–21.

Foley BJ. The role of plate tectonics-climate coupling and exposed land area in the development of habitable climates on rocky planets. The Astrophysical Journal 2015;812:36–59. https://doi.org/10.1088/0004-637X/812/1/36.

Glaser DM, Hartnett HE, Desch SJ, Unterborn CT, Anbar A, Buessecker S, Fisher T, Glaser S, Kane SR, Lisse CM, Millsaps C, Neuer S, O'Rourke JG, Santos N, Walker SI, Zolotov M. Detectability of life using oxygen on pelagic planets and water worlds. The Astrophysical Journal 2020;893. https://doi.org/10.3847/1538-4357/ab822d.

Kasting JF, Whitmire DP, Reynolds RT. Habitable zones around main sequence stars. Icarus 1993;101:108–28. https://doi.org/10.1006/icar.1993.1010.

Mulders GD, Ciesla FJ, Min M, Pascucci I. The snow line in viscous disks around low-mass stars: Implications for water delivery to terrestrial planets in the habitable zone. The Astrophysical Journal 2015;807. https://doi.org/10.1088/0004-637X/807/1/9.

Ozaki K, Reinhard CT, Tajika E. A sluggish mid-Proterozoic biosphere and its effect on Earth's redox balance. Geobiology 2019;17(1):3–11. https://doi.org/10.1111/gbi.12317.

Syverson DD, Reinhard CT, Isson TT, Holstege CJ, Katchinoff JAR, Tutolo BM, Etschmann B, Brugger J, Planavsky NJ. Nutrient supply to planetary biospheres from anoxic weathering of mafic oceanic crust. Geophysical Research Letters 2021;48. https://doi.org/10.1029/2021GL094442. e2021GL094442.

Wheat CG, McManus J, Mottl MJ, Giambalvo E. Oceanic phosphorus imbalance: Magnitude of the mid-ocean ridge flank hydrothermal sink. Geophysical Research Letters 2003;30:1895–9.

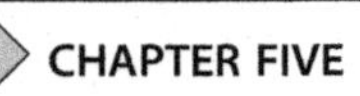

CHAPTER FIVE

Emergence of land plants and the formation of the Earth's critical zone

Introduction

For the first four billion years of Earth's existence (roughly 92% of Earth's history), the land surface was completely different from what we see today. There were no plants, no forests, no roots, and little to no soil. During that early period, important changes were occurring in the oceans, including the evolution of life itself, the subsequent proliferation of primitive marine plants and animals, and the production of oxygen and the accumulation of this critical gas in the atmosphere. But the only changes on the land surface itself were episodic intervals of ice sheet formation and the weak development and colonization of surfaces with proto-lichens and fungi. Then, in a geologic blink of an eye, a whole series of evolutionary advances began on the land about 500 million years ago.

Imagine a planet without trees, forests, or even seeds. One in which no place is shaded by overhanging branches. Where spring doesn't arrive with a burst of green overhead, and autumn doesn't bring a flaming glory of leaves? Where the literal ground you walk on is fundamentally different, with soils so thin that you might hit a rock at one shovel-depth down? Where the terrestrial water cycle was profoundly different? Well, for the first four billion years of Earth's existence, that is what you would find. That is not to say that there wasn't plenty of greenery on land—indeed, in temperate and tropical environments you likely would have been walking on a bed of moss, or at least an early relative. But trees hadn't evolved yet because they were missing one vital component that, once formed, transformed the surface of the planet profoundly, and permanently, in a geologic blink of an eye. The missing part? Deep roots. And the evolutionary development responsible for this new part? The water-bearing structure of lignin.

Plants evolved to make harder, more rigid cell structures, allowing them to reach above their neighbors to catch the sunlight. As they evolved upward, they also had to evolve downward both to capture more energy

Climate Change and Life
ISBN: 978-0-12-822568-4
https://doi.org/10.1016/B978-0-12-822568-4.00003-1

and to develop stabilizing structures so they wouldn't topple over in the wind. These structures soon became roots, reaching ever downward to mine the earth's surface for water and nutrients. Roots themselves cycle a tremendous amount of liquids and gasses between the atmosphere and the land surface, and in so doing act to dissolve the rocks around them and produce soil for the first time in four billion years. This new terrestrial strategy soon became an elevation arms race, with plants evolving into bushes, then evolving into trees. Eventually, in the late Devonian period (roughly 370 million years ago), the classic Archaeopteris (an ancient fern-like tree) forest ecosystem emerged (Algeo & Scheckler, 2010), and the Earth's surface never lost its forest cover or soils since that time. The development of soil also fundamentally changed the way weathering and erosion occur. Prior to the Devonian, wind, rain, and ice were the only agents that broke the rock into small fragments (physical weathering). After root development, the newly evolved plant acids enhanced the chemical weathering of rock into the sediment.

Once roots evolved, and plants and trees had the ability to extract deep sources of vital nutrients and water, the planet underwent its last major transition into a planet that would look, and act very familiar to us today. This evolutionary step was not without its global "pain," as it is linked to a series of biological extinction events in the ocean and a resetting of the globe's nutrient and carbon balances (Algeo & Scheckler, 2010). The Devonian hosted six marine extinction events (Fig. 5.1), including the end Devonian mass extinction, one of the "big five" with the loss of ~40% of marine families and 60% of genera (McGhee, 1996), and saw a pronounced decrease in atmospheric carbon dioxide to near contemporary levels (Lenton et al., 2016). This was not the first, nor the last, nor even the most devastating of Earth's mass extinction events, and these will be covered in much greater detail in another chapter, but it is one where we set the tone for how Earth systems have operated over the past 370 million years, and likely will for the foreseeable future.

It is through this window of time, the late Devonian Period, that we can see exactly how the Earth's biological, geological, and chemical cycles interweave in a way that reveals how important it is to view the Earth as a system. Despite various key biological innovations and planetary transitions, much is still unknown about the specific undercurrents within the Devonian ecosphere, particularly with respect to the export of plant-derived nutrients, such as phosphorus, from the terrestrial to the marine realm. While it is generally accepted that the colonization of land plants and the evolution

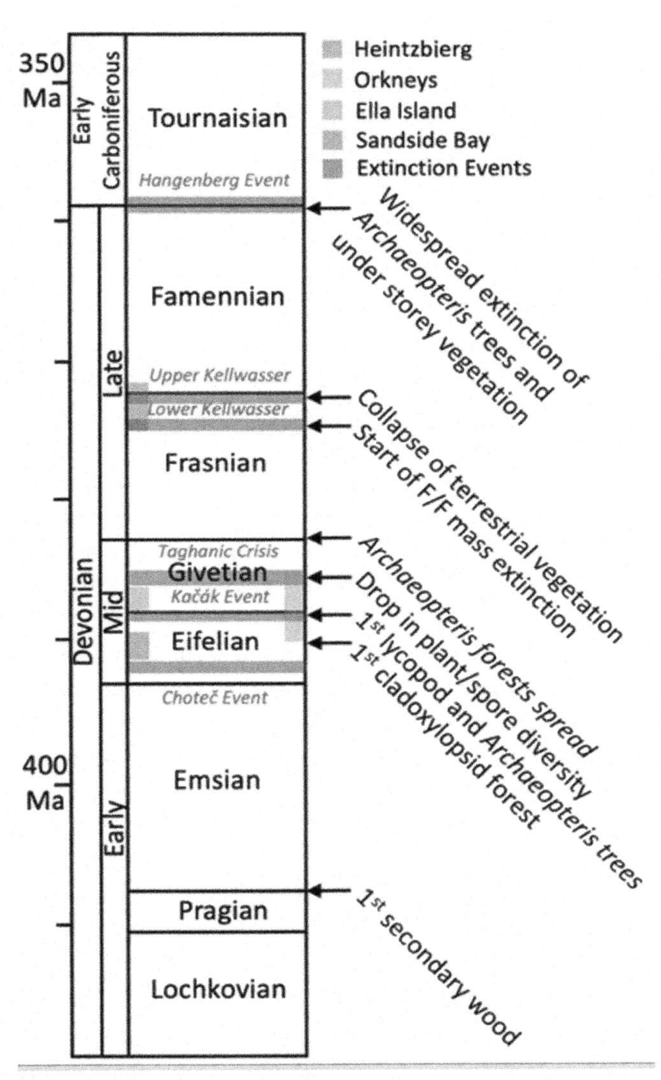

Figure 5.1 ***Major Extinction Events in the Devonian Period.*** A timeline detailing crucial events in plant development and diversity as well as key biologic crises and extinction events.

of roots would have had a marked impact on nutrient weathering, the magnitude, timing, and duration have yet to be defined. Such insights are critical to putting together all the pieces of the Devonian puzzle, particularly as terrestrial events such as the expansion and radiation of land plants have been implicated in the marine extinctions which occurred throughout the Period (Algeo & Scheckler, 1998; Becker et al., 2016).

The critical zone

You have all seen trees, and forests—their tall trunks, arching branches, green leaves and needles, and the entire array of birds and squirrels and other animals that call them home. If there is one biological unifier of all of our continents, with the exception of ice-covered Antarctica, it is the tree and the forests that they collectively make up. You can see them in your backyard or local park, from an airplane as you jet from place to place, and even from satellites orbiting far above the Earth's surface (Fig. 5.2). What you can't see from any of these perspectives is the vast, deep, and highly sophisticated system of roots under those trees and forests. Indeed, the below-ground root structure itself mirrors the above-ground structure of the tree in structure and size and typically accounts for over 10% of the total weight of the tree itself.

Tree roots have a complicated set of structures and functions, not least of which is to offset the weight of the tree canopy itself so that the entire structure doesn't tip over in the slightest wind. But the most critical components of tree roots include two main functions—a thick and deep tap root that absorbs and transports water upward against gravity, and finer roots emanating

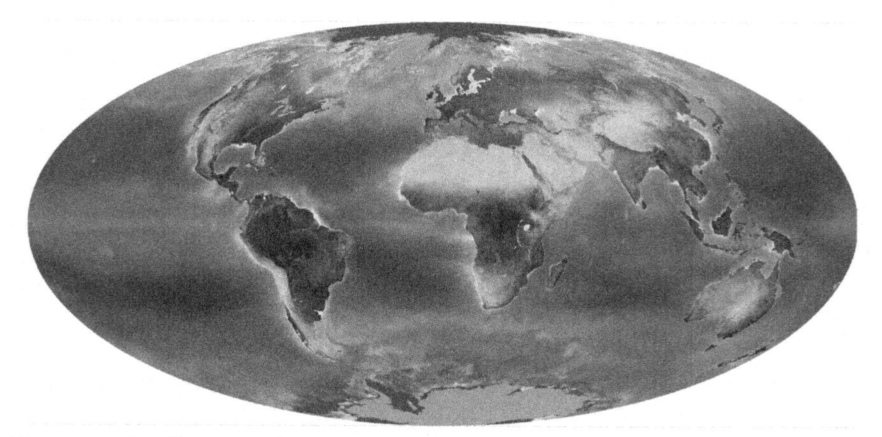

Figure 5.2 ***Satellite View of Global Plant Productivity.*** Plant productivity on land and in the ocean as imagined by NASA satellite, with areas of greatest photosynthetic activity shaded green. https://earthobservatory.nasa.gov/world-of-change/Biosphere.

from the tap root that the tree uses for water and nutrient uptake. The tap roots often penetrate deep into the soils and even the rock layers below, and often literally wedge rocks apart in their search for water and nutrients. These roots can suck up hundreds of gallons of water per day, and deliver them to the overlying leaf structure where the water is used during photosynthesis, and where the water can be transpired into the atmosphere during the process of respiration. This can be a dominant force in the hydrology and weather patterns of forests—for example, in the Amazon Basin, plant transpiration in the vast rainforest accounts for about a third of all rainfall that returns to the thirsty trees below, functioning as a vast water recycling system.

Plant roots also play a very critical role in the cycling of various critical elements on the Earth's surface. Take the element carbon for example. As discussed in Chapter 2, carbon is a great climate modulator in its role as a greenhouse gas in the atmosphere. The terrestrial biosphere is one of the key carbon exchange systems that work on this planet, alongside the marine biosphere. Terrestrial plants absorb carbon dioxide through their leaves and convert it to sugars and other organic compounds. Although the sugars are quickly metabolized by the plants themselves, and thus the carbon dioxide absorbed is released back into the atmosphere for a zero-sum-game, more complicated plants, and especially trees, use a portion of that absorbed carbon for more long-lived structural components that comprise their trunks and branches and roots. These more complex carbon compounds can persist for decades or centuries, and thus trees are also a long-term stabilizer of the carbon balance—which speaks to the worry that climate scientists have about how rapid deforestation and burning can rapidly convert that long-lived carbon storage into carbon dioxide, which worsens climate change. Furthermore, some of the metabolism of plant sugars occurs through roots, where the released carbon dioxide can mix with soil water to become carbonic acid, increasing the dissolution of minerals in soils.

As noted later in this chapter plants, and particularly their roots, strongly affect the balance of plant-essential nutrients such as nitrogen and phosphorus. Collectively, then, the activity of plants on the modern Earth creates a soiled envelope of the Earth's surface, which strongly impacts the hydrology, chemistry, and climate of the planet. This envelope, in which rapid geological, chemical and biological reactions on land occur is what is called the "critical zone" (Fig. 5.3). This zone is the interface between the atmosphere and the lithosphere, which is the term to describe Earth's crust. On modern Earth, the critical zone is largely defined by the layer of soil between

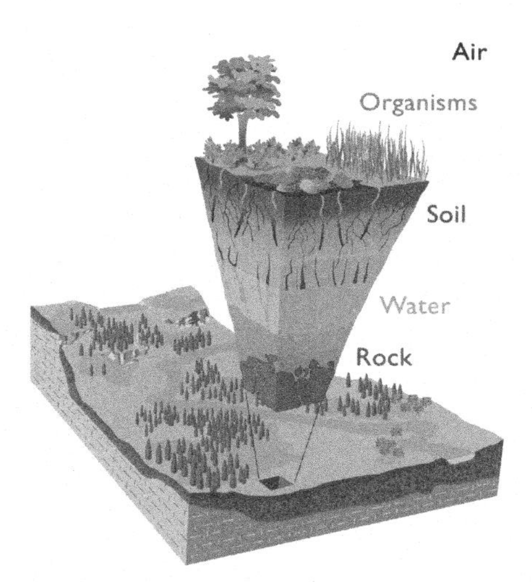

Figure 5.3 ***Earth's "Critical Zone."*** An exploded view of the organisms, soil, and water that comprise the critical zone—the surface layer separating the atmosphere from the bedrock crust of the planet. *From Chorover, J., Kretzschmar, R., Garcia-Pichel, F., & Sparks, D. (2007). Soil biogeochemical processes in the critical zone.* Elements, *3, 321—326. https://doi.org/10.2113/gselements.3.5.321.*

the atmosphere and hard, unweathered rocks below. The soil layer can be thin or nonexistent, in places where wind strips it away, or topography is so steep that it never accumulates and instead is deposited in lowlands. It can also be nonexistent on surfaces that are freshly exposed when ice or water has recently retreated, or in environments where new material is freshly added, such as flowing lava or ashfall in Hawaii. But it can also be quite thick, reaching depths of 100 m or more in low-lying, deeply weathered environments such as found in the Amazon Basin. But what is the origin of soil—how does it form?

Soil forms from a combination of the mechanical, or physical, breakdown of rocks by water, wind, or ice into smaller pieces, and by chemical reactions with water that dissolve or otherwise transform the minerals that comprise the eroded rock. A mountain slope may experience a significant amount of physical weathering by wind, water, and ice, but the product of this weathering is swept down into the mountain valleys due to gravity, not spending much time on the mountain slope itself before this erosion. Because chemical weathering reactions between water and minerals take some time, this material might appear very "fresh" to a geologist's eye, who can easily identify the minerals that make up this eroded material.

But once in the low valley, further erosion and transport are slowed down tremendously, and the eroded material now can benefit from time in that environment for rainwater and moisture to begin transforming the minerals. In this way, through physical deposition and chemical alteration, a soil can form. In this case, the soil may be forming mostly above the bedrock, from the continual addition of new material from above.

But some soils form instead of the Hawaiian lava-type scenario. Here, a fresh layer of basaltic lava might become deposited in a valley and solidify. Soon, plant seeds that are transported via wind might land on the lava, and if rain comes along, those seeds will sprout (germinate), and the fledgling roots will start burrowing down into the rock, first to stabilize the young plant in place and second to begin breaking the minerals apart that form the rock, seeking critical growth nutrients. As this process proceeds, the lava becomes even more hospitable to plant seed germination because the job has been made easier by the plant pioneers who helped "soften" up the lava surface, which is now covered with a thin layer of softer mineral material that is easier to burrow in to and is starting to be able to hold water. As this scenario plays out, the soils become deeper and deeper and begin exhibiting a set of layers associated with a more mature soil and reflecting the varying degree of chemical reactions and plant litter deposition that occurs over time. In this case, the soil is forming into the bedrock, in a physical and chemical process that begins at the surface and proceeds to depth.

In reality, though, most soils actually are a product of both the processes outlined above—made from material deposited from above but also including a soil-forming process that involves the transformation of the bedrock below. These soils represent the foundation upon which human society is maintained, and upon which all modern terrestrial ecosystems are built. But this landscape was not always there—rapid evolutionary developments over 370 million years ago resulted in our modern planetary surface, and in so doing likely triggered a series of global extinction events.

Lignin, water, and roots

The process of biological evolution is not intentional, nor even a continuing progression of species improvement. It is simply dumb luck—if a random mutation makes an organism more successful in surviving to the point of reproduction than one without a mutation, then those genes will pass along and alter later generations to in turn be more successful at passing along this mutation. If the latter occurs, i.e., a mutation makes an organism less successful at reaching reproductive maturity, then that mutation will die out. Additionally, if an organismal line is shaped by a set of mutations that are advantageous for a given environment and that environment changes, then those mutations are no longer strategically beneficial to that organismal line. If the organism can't move or otherwise have a range migration, then it will be less reproductively successful and will eventually die off. If the environment changes everywhere, then of course there is no hope for that mutation.

In light of this, how did evolution shape the amazing forest transformation of the planet 370 million years ago? It actually started earlier than that, likely as much as 400 million years ago or earlier, with the evolution of lignin. Lignin forms the stiff support tissues of trees and acts as a water conduit in stems. Lignin is rigid, which in the Devonian lent organisms that produced it in a mutation the advantage of height and depth. With height, a lignin-enabled organism can build its stems in such a way that they don't just lay on the ground and photosynthesize on the ground, but rather can rise up a meter or two, to capture sunlight before it hits the ground, effectively shading out the ground-bound organism that didn't produce lignin. So this lignin-enabled organism would be more likely to grow to maturity and reproduce and propagate the lignin mutation.

Height has the advantage of getting a plant more sunlight and beating out more diminutive competitors for nutrients, to be sure, but it has the disadvantage of requiring much better water conduction, which has to work against gravity. It also could produce a structure that is "top-heavy" making the plant likely to topple in the face of heavy rain or wind. Finally, the structure of taller plants requires a tremendous amount of resource investment, with very little in the way of direct return. A plant has to make all of its own food from photosynthesis, before then consuming that food in a process called respiration—animals only do the latter, relying on other organisms to capture the Sun's energy. In many plants, and in all trees, the machinery to perform photosynthesis is in the leaves or needles. But think of a normal tree—what do you see? Sure, leaves and needles up top, but also branches, a trunk, bark. All of these structural components are necessary for a tree to stand high and protected, and capture as much sunlight as possible, but none of them actually perform photosynthesis—they are all support systems.

These support systems require a lot of resources, which have to come from somewhere. And that somewhere is largely the root systems of the trees, which supply the tremendous water demand that trees have but also are conduits for the trees to tap into the vast bank of soil nutrients, such as phosphorus and nitrogen, which are critical to all functions of a tree. For example, phosphorus is the key ingredient in the most common biological molecule on the planet—adenosine triphosphate, or ATP. ATP is the energy storage system for photosynthesis and thus is crucial for plants. Additionally, nitrogen is key to the construction of many of the key compounds in organic matter, such as proteins. For both of these nutrients, the only source is the soil themselves, and the only supply system is rooted. Modern soil systems have bolstered the ability of plants to "mine" the soil for nutrients. In the case of phosphorus, there is an entire soil ecosystem that is based on fungi that help plants drag phosphorus to roots for uptake. For nitrogen, specialized bacteria coexist with some types of plants Fig. 5.4 that can turn nitrogen gas in soil pockets and water into the nitrogen form that plants need.

For the evolving trees of the Devonian, roots played three roles—to provide a stable base so that trees could grow high, to meet the huge nutrient demands that this new, complicated organism required, and to supply vast amounts of water required for photosynthesis.

The evolution of Devonian forests

It might be convenient to assume that the evolution of roots and trunks would yield permanent forests of trees starting in the later Devonian. But as with many evolutionary developments in the past, this brand new way

Figure 5.4 ***The Root Nodules of a 4-Week-Old Medicago Italica.*** The root nodules of a 4-week-old Medicago Italica inoculated with Sinorhizobium meliloti. These root nodules have the capability to capture nitrogen gas and transform it into the plant nutrient nitrate.

of functioning went through a series of tumultuous changes until true, permanent forests expanded to cover the world. Before the middle of the Devonian, land plants were confined to areas immediately adjacent to bodies of water, had limited root systems, and had yet to develop the necessary biological innovations that would lead to arborescence, the development of the first trees (Algeo & Scheckler, 1998, 2010). Continental interiors would have been bereft of significant vegetation. By the Middle to Late Devonian, land plants began to diversify and spread into continental interiors and uplands, populating a once barren landscape with vegetation (Piombino, 2016). The diversification and expansion of plants led to the first appearance of trees and the widespread propensity of plants to produce seeds.

The earliest of these trees that appeared was *Archaeopteris*, a large progymnosperm that grew up to 30 m in height (Stein et al., 2020). One of the most fascinating aspects of this first forest-forming tree is that it had a broad canopy to effectively capture light and shade out the competition, while also allowing light to penetrate to lower branches and leaves. The tree itself looks much like a modern fern with fronds (Fig. 5.5), but was quite tall, like a top-heavy Christmas tree (Fig. 5.6). Although not a seed-bearing plant, *Archaeopteris* was the dominant large plant until well into the Late Devonian, comprising forests that spanned from the equator to the very high latitudes. It was later displaced by more advanced seed-bearing plants (Stein et al., 2020).

But even with the apparent dominance of Archaeopteris forests, a number of other relatively short-lived tree groups evolved during this time and can be seen in the fossil record. One, called Cladoxylopsid, is thought to be an ancestor of the modern ferns and horsetails, and as a tree had little in the way of foliage. The trees were palm-like (Fig. 5.7), but instead of leaves had branched twig-like appendages, which presumably had a photosynthetic function in the carbon dioxide-rich atmosphere of the Devonian (more on that later). The Cladoxylopsid didn't have the more modern tree-like vascular plumbing system with growth rings growing outward but instead had a relatively complicated set of fibrous strands, which allowed the tree trunks to grow and expand from within the trunk (Berry, 2017). It also did not produce seeds like other tree groups, which along with its relatively complicated and inefficient growth dynamics led this group to go extinct toward the end of the Devonian. Fossils from this group are largely just the stems and roots (Fig. 5.8), so it is difficult to determine foliage shapes or patterns.

During this long interval of evolutionary experimentation, with different tree groups and forest ecosystems appearing and disappearing across the

Figure 5.5 ***Fossil Archaeopteris Fronds.*** Fossil fronds from the first major forest-forming tree, the Archaeopteris. From the Waterloo Farm lagerstätte in the Eastern Cape, South Africa. These represent the only high latitude species of Archaeopteris as yet described. *By Funyu123 - Own work, CC BY-SA 4.0, https://commons.wikimedia.org/w/index.php?curid=89841171.*

landscape, one major component was constant—these trees had roots, and these root systems were quite extensive. The earliest evidence of roots associated with vascular land plants appeared in the Late Pragian, or Early Devonian (Algeo et al., 2001). More substantial root systems did not appear until the Middle to Late Devonian and were notably associated with *Archaeopteris* (Algeo et al., 2001; Kenrick & Strullu-Derrien, 2014). Significant root development led to the creation of modern soil weathering processes, nutrient cycling, and extensive soil development (Algeo & Scheckler, 1998; Kenrick and Strullu-Derrien, 2014; Morris et al., 2015). *Archaeopteris* were likely long-lived and thus established significant soil profiles (Algeo & Scheckler, 1998) (Fig. 5.9) with a measurable impact on local biogeochemical cycling

Figure 5.6 ***Archaeopteris Reconstruction.*** Reconstruction of the possible growth habit of the Archaeopteris tree. *From Falconaumanni, CC BY-SA 3.0, via Wikimedia Commons. https://commons.wikimedia.org/wiki/File:Archaeopteris_reconstruccion.jpg.*

and nutrient loads into the oceans. Indeed, many studies have shown that roots dramatically alter soil formation, and thus soil biogeochemistry (Filippelli, 2002, 2008). The development of extensive root systems, stress-resistant seeds, and an overall increase in size subsequently contributed to the expansion of land plants into continental interiors and uplands.

Impact of roots and forests on nutrient cycles

Phosphorus is a critical biolimiting nutrient on land and in aquatic systems (Filippelli, 2008, 2010). Unlike other essential nutrients such as nitrogen and carbon, phosphorus has no natural stable gaseous form. Phosphorus can be transported long distances through the suspension of dust particles in the atmosphere, however, this delivery method constitutes a minute fraction of the total phosphorus cycle (Filippelli, 2008). The only significant source of phosphorus to the oceans and other bodies of water is through the weathering of continental materials and ultimate delivery by way of rivers (Fig. 5.10) (Filippelli, 2002).

Figure 5.7 ***Cladoxylopsid Forest Reconstruction.*** An artist's impression of a fossil forest.

Figure 5.8 ***A Fossil Soil Layer, Called a Paleosol, Containing Imprints of Devonian Tree Roots.*** Deep and thick roots capable of extracting water and nutrients from the soil and deeply weathering bedrock first evolved during the Devonian Period. *By Eduard Solà - Own work, CC BY-SA 3.0, https://commons.wikimedia.org/w/index.php?curid=3306832.*

Figure 5.9 ***Modern Soil Profile.*** A cut-away view of what modern soil looks like, including a darker, more organic-rich at the top, a red iron-rich soil layer, and a thick, tan, chemically depleted layer on the bottom. *CC BY-SA 3.0, via Wikimedia Commons.*

Soils are the ultimate chemical alteration factory for phosphorus. Imagine a freshly erupted lava covering a valley. By nature of being all rock, there is no soil at all there, and all of the phosphorus that is present is there as mineral phosphorus—typically in form of a mineral called "apatite." As time goes on, the rain will start reacting with some of the surface layers of the lava, and small lichen and other young plants will colonize this fresh surface, further dissolving the lava as their roots begin to penetrate the rock. As soil development proceeds, the phosphorus originally there as the mineral apatite (which is quite easy for plant roots to dissolve and for the plants to take up) gets transformed. A fair amount of it might be washed away with rainwater as the plant roots penetrate deeper and deeper into the former lava. But that which isn't washed away begins to transform into other chemical forms, including one where the phosphorus is bound up with newly-formed iron minerals, called "occluded" or hidden phosphorus because roots can't get at this phosphorus very quickly (Fig. 5.11). Additionally, as the plant grows leaves and branches, and as these and other parts of the plant fall to the ground, the phosphorus that was incorporated as a key growth

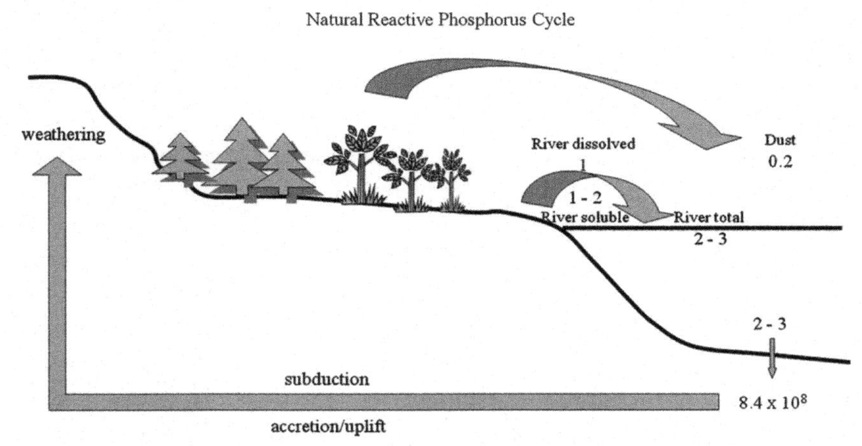

Figure 5.10 ***The Phosphorus Cycle.*** The natural cycle of reactive phosphorus (P), with phosphorus input to the oceans from riverine and dust sources, output via sedimentation, and recycling via tectonics (fiuxes in Tg P/year, sedimentary reservoir in Tg P).

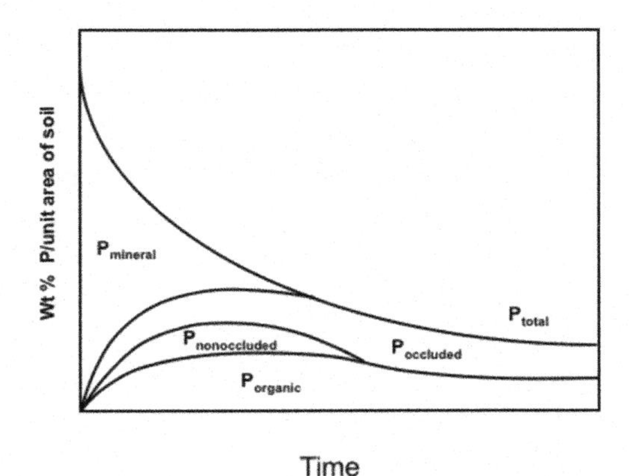

Figure 5.11 ***Changes in Soil Phosphorus Over Time.*** Upon the development of soil on a fresh rock surface, phosphorus undergoes a number of geochemical transformations. These transformations involve phosphorus being converted from the mineral form to less plant-available forms, such as occluded (bound to iron minerals), organic (as biological debris), and nonoccluded (bound to the surface of minerals). Note the consistent loss of total phosphorus from the soil profile over time.

nutrient is now stored as an organic-bound form of phosphorus. Finally, some of the phosphorus will remain, very loosely bound to clay minerals that have formed as the soil matures. In this way, the birth of soil takes the original mineral phosphorus and transforms it into a host of other phosphorus compounds, losing a fair proportion of the phosphorus along the way to run-off. It is thus expected that a landscape with little to no vegetation would transition through these phases, beginning with predominantly mineral sources of phosphorus and shifting to more occluded sources. Likewise, it would be expected that the overall release of phosphorus in a landscape would peak very quickly upon plant colonization and stabilize over time (Bruckner, 2017; Filippelli & Souch, 1999).

The development of soil fundamentally changed the way weathering and erosion occur. Prior to the Devonian, wind, rain, and ice were the only agents that broke the rock into small fragments (physical weathering). After root development, the newly evolved plant acids enhanced the chemical weathering of rock into the sediment. The ramifications of this massive transformation of the earth's surface have been explored from several fronts, including soil formation studies, changes in ocean chemistry, and even mass extinctions. Indeed, fingers have been pointed at the soil formation process and the resultant loss of phosphorus from land during the Devonian as a trigger for global mass extinctions in the oceans, through a fertilization effect. Briefly, modern soils are markedly low in nutrient phosphorus, due to a loss of this element during soil formation processes and transport into waterways and eventually the ocean. Once in the ocean, this nutrient fertilizes algae and spurs excessive algal growth. Indeed, overapplication of phosphorus in fertilizers in the Midwest is the distal cause of the Gulf of Mexico Dead Zone, a region where phosphorus lost by fertilizer and soil run-off from farm fields fuels algal blooms (Bruckner, 2017). These algal blooms in turn support a host of algae consumers who suck all of the oxygen out of the surrounding water during this process, making the region deadly to fish and shellfish. In a similar fashion, some researchers have conjectured that the "mining" of phosphorus during the expansion of land plants and the development of soil during the Devonian resulted in a similar Dead Zone, but this one was worldwide and well-recorded in the geologic record.

Marine anoxia and mass extinctions

Geologists have uncovered a series of four minor to moderate evolutionary setbacks, or mini-extinctions, throughout the Devonian Period, which collectively is considered a Mass Extinction interval. Approximately 80% of Devonian animal species went extinct during this time, exclusively in the marine environment. Interestingly, at the same time on land, there were major evolutionary events related to forest species as described above but also the first emergence of tetrapods, animals that walked on all four limbs on land. Thus, the Devonian is a paradoxical period where the marine environment suffered repeated devastation but the terrestrial one had significant biological innovation. As described at length in Chapter 6, mass extinction events are scattered throughout the geologic record, and most seem to be triggered by some major geologic event, such as massive volcanic eruptions or asteroid impacts, which then strongly impacts the global climate and the stability of ecosystems. The resulting extinction events tend to affect both terrestrial and marine species, though not always in equal measure. The Devonian extinction was different, as there is no strong evidence for volcanic or impact triggers (although a recent study pointed to a complicated mechanism involving the short-term loss of our protective stratospheric ozone layer). The strongest evidence for the Devonian extinction mechanisms lies in the extremely thick, organic-rich layered shales on the seafloor, and now found outcropping throughout the eastern portion of the USA, Greenland, Scotland, and many other areas that were once part of a vast emergent ocean basin.

Sedimentary rock deposits from ancient oceans are typically the best record of geologic history because they have the capacity in a certain environment to accumulate in layers year after year, and millenia after millenia. These layers reflect what was occurring in the ocean at the time they were deposited, and, through a certain degree of inference, record conditions on land as well. Much like tree rings can record past wet or dry spells on land, marine layers can reflect times when the ocean was very productive, with fish and algae and other organisms thriving, and times when conditions were harsh. One of the key clues to what drove the mass extinction events during the Devonian Period are vast sequences of layered shale deposits that were formed under conditions of very little, or no, dissolved oxygen in the ocean. Geochemists and sedimentologists can use a host of chemical and textural evidence to determine the relative amount of oxygen in water, and the prognosis for the Devonian was not good.

Oxygen gets into ocean water by diffusion from the atmosphere at the ocean's surface. The amount of dissolved oxygen on the surface depends on the concentration in the atmosphere, and the temperature of the water—warm water tends to hold less dissolved gasses than cold water, a science lesson intuitively understood by anyone who has left a carbonated soda out of the refrigerator too long just to return to a flat syrupy disappointment. But the exchange between the ocean and the atmosphere only occurs at the ocean surface, and in general, the top 200 m or so of ocean water mixes up with the atmosphere on a regular basis, meaning that the top meter of the ocean has about the same oxygen as the layer 200 m down. The oceans, however, have an average depth of about 3800 m, and most of that deeper water is quite old, having mixed downward from the surface centuries or millennia before, and no longer exchanging freely with the atmosphere. In those old, dark ocean depths, some significant changes can occur in water chemistry, particularly for oxygen.

Oxygen is a product of photosynthesis, and marine plants produce oxygen in the surface ocean all the time. That, combined with frequent mixing of the surface ocean, yields relatively high oxygen in these waters. Animals need oxygen to survive because, unlike photosynthetic plants, animals only metabolize sugars and consume oxygen in the process, and thus the ocean is filled with fish and clams, and crabs. But as one goes deeper into the ocean, that oxygen starts to decrease, because microbes and animals are consuming the constant rain of organic matter that falls from the surface and consuming oxygen in the process, with no return of that oxygen from mixing with the atmosphere or from photosynthesis (it is simply too dark down there). In most of the modern deeper ocean waters there is enough oxygen for animals to flourish, but in some isolated areas where excess plant production has caused high rates of organic rain to the depths, nearly all, or sometimes absolutely all of the dissolved oxygen is removed from the water. At that point, animals can't survive, and animals that can't move or migrate quickly die. Additionally, in areas without dissolved oxygen other noxious chemicals can be released into the water, like hydrogen sulfide with is a potent poison for most animals.

Returning back to the Devonian, the oxygen-less layers in marine sediments cover the entire ocean basin in some cases. It is then no surprise to find the repeated final occurrence of a number of marine species during this time, as organisms die in poisonous seas. These oxygen-poor seas may have been the result of several factors, including an ocean system that doesn't circulate much at all, with deeper waters staying stagnant for tens of thousands of years. In that scenario, there is perhaps enough time for even a normal amount of organic matter raining down in the deep ocean to support enough consumption to remove the oxygen. But a similar driver could be from excess fertilization of the oceans during this interval—not from Devonian farmers overapplying nitrate and phosphate to their crops, but instead from the extremely rapid release of large amounts of phosphorus during this singular moment in Earth's development of plant roots and soils.

Compounding this input might have been another geochemical process occurring from the ocean depths themselves.

Phosphorus tends to absorb onto iron particles in the ocean and in sediments, thus removing phosphorus from the use by plants. But in low oxygen and no oxygen conditions, the iron particles dissolve, and release phosphorus back to the water. There it can once again feed marine algae, which would be consumed by an animal further removing oxygen from the water and sediment, and so on, and so forth (Algeo & Scheckler, 1998). This runaway train of oxygen loss would sustain ocean anoxia and mass extinctions for some time even after the excess phosphorus input from root and tree evolution slowed down or stopped (Algeo & Scheckler, 2010).

So what caused the Devonian mass extinctions? Excess phosphorus input from tree evolution? Ocean stagnation? Both? Neither? To explore this question, the best place to start is with phosphorus, and the most appropriate question to ask is if the evolution of trees and roots caused enough phosphorus release from the terrestrial landscape, for long enough, over an area great enough, to have driven the overfertilization and marine anoxia clearly observed in the geologic record? And the evidence seems to say "maybe" (Fields et al., 2020).

So, was phosphorus the culprit?

Shifts in the phosphorus balance have been implicated as a driver of some of the Devonian marine crises but without direct evidence. Nearly all of the marine extinction events in the Devonian were characterized by widespread anoxia (Becker et al., 2016). As noted above, for these vast ocean anoxic zones to form, it is reasonable to hypothesize an external fertilization mechanism, much as we see in the modern Gulf of Mexico driven by the wash off of excess agricultural fertilizers including phosphorus. Phosphorus is mobilized rapidly in young landscapes and landscapes that are rapidly colonized by land plants. Not yet tested in this hypothesis is whether a nutrient pulse (or pulses) of sufficient size and on a global, or even basin-wide scale, to drive these significant marine changes occurred. This is particularly important as these marine anoxia events persisted over millions of years, whereas the lifetime of a molecule of phosphate in the ocean is only about 20,000 years, so about 100 times shorter than needed. Of course, the lifetime of the phosphate might have been longer in the Devonian system particularly as more recycling was occurring in the anoxic water and sediments, but it would require MUCH more recycling for this alone to be plausible.

But instead of a single pulse, might multiple pulses of weathering-induced phosphate from the land be enough to drive the anoxic patterns? Perhaps, if they are of adequate scale and frequency. New evidence is emerging on the dynamics of the global Devonian phosphorus balance from an unlikely source—large lake systems that persisted through this time which served as "transit systems" for the material being weathered from the continents and then transported to the ocean. During this time of the Devonian large lake, basins were being formed by geologic activities that, at least temporarily, stretched across large portions of the Devonian subtropics. These deposits are a portion of a larger system generally called the "Old Red Sandstone," to distinguish them from the "New Red Sandstone," a typically practical yet uncreative solution employed by British geologists of the day. They comprise a full series of rock types and depositional environments, of which red sandstones are one part but also lakes and estuaries.

Lakes have proven to be excellent recorders of terrestrial processes, including the transformations of phosphorus in local soils discussed earlier as well as the degree of phosphate weathering and release that occurs over time. For example, sediment cores recovered from small glacially-impacted lakes reveal the rapid shift in soil development and phosphorus chemistry that would also be apparent on fresh lava flows if one had several thousand of years to study them—lakes show the process when it happened. As glacial activity ceased at the end of the Last Ice Age, these ice-scraped landscapes began weathering and forming soils and were inhabited by plants. This process occurred quickly and basically took the landscape from bare rock to mature soil in 10,000 years. In so doing, the expected transformations of soil phosphorus chemistry were observed, as well as the significant loss of phosphorus from the system during the period of rapid soil development (Fig. 5.12). The phosphorus loss was largely just in the transition interval, and on a grander scale would be exactly that type of phosphorus "pulse" that would have been released by the evolution of roots and trees and forests during the Devonian. This would be a short, one-time deal unless the landscape was again disturbed by climate or erosion or tree extinction, all of which might "reset" the system for another round of soil development.

But is there any evidence of this phosphorus release pulse during the Devonian, and was it enough phosphorus to kick the marine system into anoxia? The answers are yes, and maybe. A reconstruction of the rate of phosphorus release from land during one interval in the Devonian seems to support this concept (Fig. 5.13). This sequence was from lake sediments,

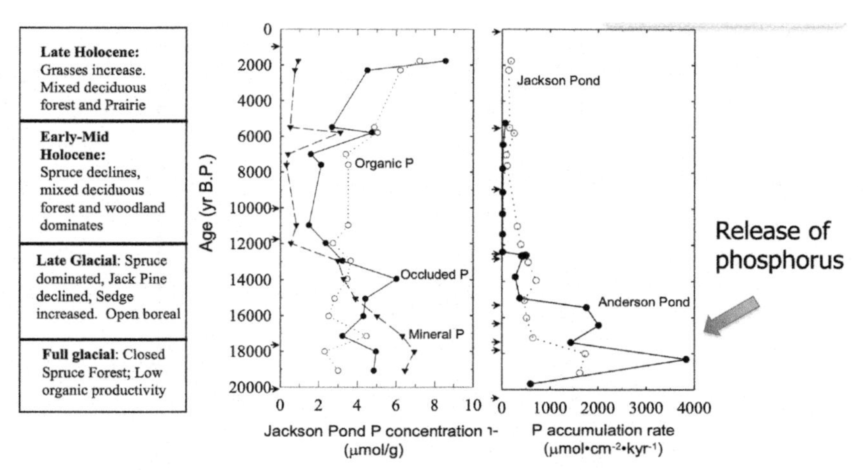

Figure 5.12 ***Lake Sediment Record Showing Changes in Phosphorus geochemistry and Release During the Development of Soil.*** Lake sediment core reveals changes in soil phosphorus chemistry after recovery from glacial to nonglacial conditions. Note the initial release of a substantial amount of phosphorus from the landscape during the transition and the initiation of soil development alongside ecosystem maturation.

and the interval measured was associated with terrestrial evolution and a major marine anoxic event and consequent marine extinction (Smart et al., 2019, 2022). Although the time interval is not known as well as for the recent glacial example above, enough is known to be able to determine the nature of phosphorus release and the general amount released. Both of these show that it is quite similar to the postglacial analogy, in terms both of how brief the pulse was and most importantly how much phosphorus was released. Although only one record, from a very long time ago, it is the first such view of soil development and phosphorus release caught "in action" for the Devonian and is consistent with the hypothesis that the evolution of plant roots caused significant perturbations to the terrestrial phosphorus balance.

But is this pulse of a magnitude and duration to fertilize the ocean with enough phosphorus to cause anoxia? To address that, one needs to turn to models, which are our best way of understanding how realistic some our assumptions are. The models are relatively complex and based on our best understanding of the relationship between phosphorus, marine productivity, and anoxia. Using the rate of phosphorus input from Fig. 5.13 and assuming 10% of the land area of the Devonian released this much phosphorus at the same time from the rapid evolution of a forest ecosystem, that phosphorus

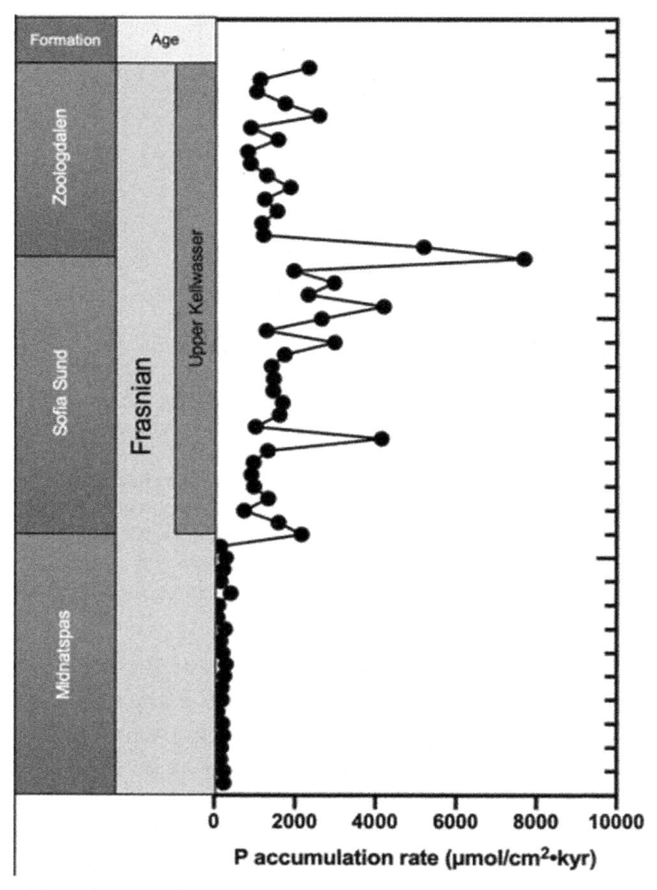

Figure 5.13 ***Phosphorus Release Rate From Land During the Devonian Period.*** The rate of release was calculated based on the phosphorus concentration and estimated sedimentation rate for a fossil lake sequence in the Devonian Period.

would stimulate enough marine biological productivity to turn the ocean largely anoxic (Fig. 5.14, Smart et al., 2019). So in that sense, the actual data supports the basic aspect of the hypothesis that forest evolution can result in marine anoxia, which in turn could cause a marine extinction event. But note the critical piece here—that input of phosphorus is short-lived, and its impact on marine anoxia is similarly short-lived. Within 100 thousand years or so the ocean returned to its previous, more oxygen-rich state. This phosphorus pulse event would have to be repeated a number of times in order to support ocean conditions that would be regularly anoxic. This

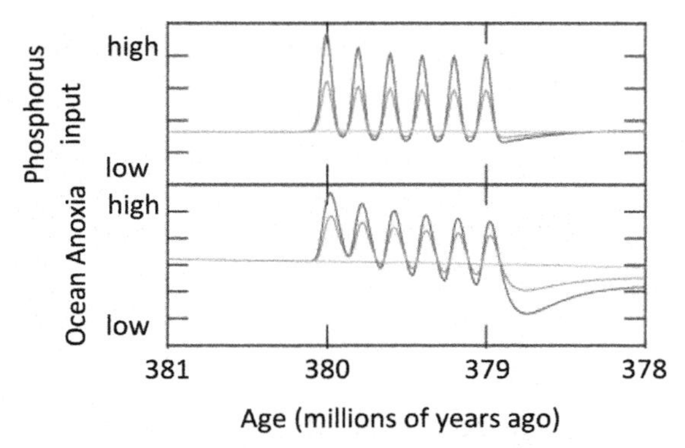

Figure 5.14 ***Global Phosphorus Input to the Ocean and Extent of Ocean Anoxia Modeled for an Interval of the Devonian Period.*** Model of the degree of ocean anoxia predicted from weathering pulses of phosphorus matching repeated episodes of those determined from the Devonian lake record shown in Fig. 7.13.

scenario is supported to some extent by the fossil forest record during this time, and the repeatedly anoxic ocean would likely be enough to drive significant marine extinction because species would not be able to recover adequately during an oxygen-rich phase before a new anoxic wave roared into the ocean.

One frustrating thing about geoscience is that many assumptions need to be made, and the few "facts" that we have (i.e., geological records) have to be viewed through the very long and often distorted lens of geologic time. It is true that those assumptions are often based on how the modern Earth works, and thus are based on some reality. For example, the suggestion that phosphorus is released at a high rate once a landscape is colonized by plants and develops soils is supported by multiple recent records. And the fact that a high amount of phosphorus released into the ocean causes zones of anoxia is supported by direct observations from the Gulf of Mexico. Additionally, the fact that the Devonian ocean underwent long periods of anoxia and consequent mass extinctions is supported by the vast deposits of Devonian-age rocks that show clear signs of being deposited at the bottom of an anoxic ocean. But making the connection, that root evolution caused phosphorus release from land, which in turn fertilized the ocean, which in turn led the ocean to become an anoxic and deadly habitat for marine animals, is always fraught with uncertainties. But along with the frustration of not knowing these connections for sure comes the thrill of the mental experiment of making these connections!

One final possibility for Devonian extinctions—climate change

Beyond phosphorus, ocean stagnation, and even asteroid impacts or volcanism, remains one other phenomenon that could be linked to one or more of the Devonian extinctions—climate change. Consider how the balance of one of the critical greenhouse gas components, carbon, changed from the various processes described above. First, the growth of a robust terrestrial forest cover and the birth of the "critical" zone both would have pulled carbon from the atmosphere and sequestered it in plants and soil organic matter. Since soil organic matter, in particular, is very long-lived (hundreds to thousands of years), this transfer of carbon from the atmosphere to the land surface would have decreased the concentration of carbon dioxide in the atmosphere. Second, consider the massive amount of organic matter, and associated carbon, that was deposited on the ocean bottom during Devonian anoxic events. This also ultimately represents carbon being removed from the atmosphere by marine plants and transferred to the seafloor. This carbon in particular would not likely be recycled and returned to the atmosphere other than through the slow geologic processes of plate tectonics (tens of millions of years). Finally, the evolution of plant roots and the increase in weathering of rocks also pull carbon dioxide out of the atmosphere—the carbon is converted into a dissolved carbonate molecule that is transported to the ocean and constitutes a building block of animal shells and coral skeletons, which like organic matter on the seafloor only returns to the atmosphere on the very slow geologic timescales of plate tectonics.

It seems that atmospheric carbon dioxide is being hit from three sides, namely trees, sediments, and weathering, and all at the same time. All of these processes are pushing carbon dioxide in the same direction—out of the atmosphere and onto the surface. The expected result would be a colder climate, but is there evidence for this in the geologic record, and would it be enough to cause some of the observed extinction events? To place the Devonian in a climate context, it is important to note that the amount of atmospheric carbon dioxide was roughly seven times the current amount, leading to global temperatures of about 17°C, compared to today's average of about 14°C (Fig. 5.15, Smart et al., 2019). The modeled drop in Devonian carbon dioxide from the various carbon removal processes outlined above was from seven to about five times present values, with a consequent temperature drop of about 2–3 degrees, so close to our current temperature.

This is a substantial drop in temperature and may have been implicated in the formation of glacial ice. There is some evidence of glaciation from rock deposits in modern-day Brazil (which at that time was located near the South Pole and thus quite prone to potential glaciation), but it is difficult to determine the global extent of glaciation. If enough water was removed via land-based ice formation, it is possible that the sea level dropped and thus limited the potential habitat for Devonian reef systems, which did indeed suffer major extinctions at this time. But many other marine species, including mobile ones, went extinct at this time, and it is hard to imagine how these would have been substantially impacted by a temperate drop given that they are largely buffered by the ocean and by their own ability to migrate to more tropic areas. Also, some researchers point to a potential *increase* in temperature at the end of the Devonian Period and think that this caused a string of events that eroded our protective stratospheric ozone layer and exposed terrestrial life to considerable amounts of harmful UV radiation (Marshall, 2020). Or even, as some scientists claim, a nearby supernova that burned off our protective ozone layer and fried the planet (Fields et al., 2020). But again, this would not have affected marine organisms to a great extent because the ocean absorbs most of this harmful radiation. Regardless, climate change might have just been one more nail in the coffin for many of the Devonian ecosystems already rocked by marine anoxia.

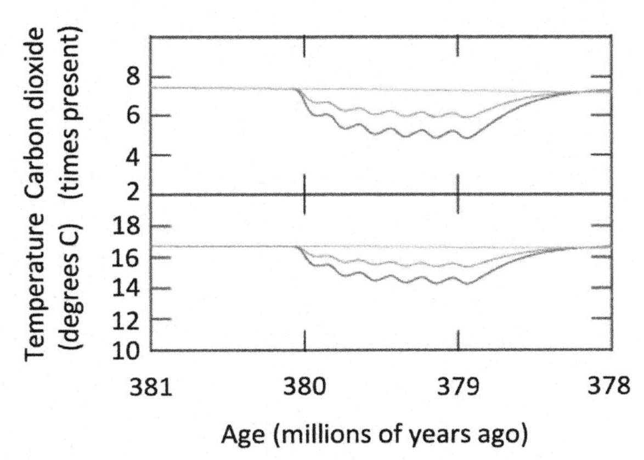

Figure 5.15 ***Atmospheric Carbon dioxide Concentrations and Global Temperatures Modeled for an Interval of the Devonian Period.*** Model of the degree of atmospheric carbon dioxide and resultant global temperature from repeated episodes of phosphate input to the ocean presented in Fig. 5.14.

Summary

The singular transformation of our terrestrial surface in the Devonian Period due to the evolution of roots in plants remains one of the most profound transitions in Earth's history. No question that other such transformations exist, such as the global rusting of the planet from cyanobacteria, the emergence of animals from the oceans onto land, the shocking demise of the dinosaurs, and the rise of the mammals in their place. But consider again the post-Devonian land surface and its Critical Zone, which then becomes the filter between atmospheric and geologic processes on land. This filter holds a tremendous amount of water, and organic matter, and is a weathering factory. Consider towering redwood trees and the dense Amazon tropical rainforest. Even the oak and maple trees lining the streets of towns in Indiana, providing shade for children to play on a hot summer day. This "root revolution" made possible by the evolution of lignin and the formation of vascular structures in plants was really a marvel in biology and in engineering. This interval also provides profound lessons about how intertwined the various cycles are on Earth, and how resilient Earth systems are to change. Sure, if you are a species that went extinct in the Devonian due to these changes, you would be none too happy. But soon after your demise, you would find your place on this planet, your niche, filled by another species, better adapted to the "new normal" and ready to carry on. A humbling reminder, even to we humans, that on this Earth, life goes on, and no species is too special to be immune to the forces of nature that cause extinctions.

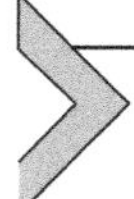

Did you know that?

Giant insects ruled the planet after the Devonian

Well, "ruled the planet" might be a bit of a stretch, but the post-Devonian period brought an age of insect gigantism. After the Devonian Period was the Carboniferous Period, appropriately named after the vast amount of organic carbon buried in swamps and marine system at this time. Many of the coal deposits in the USA hearken back to this time. As discussed in an earlier chapter, when you remove more carbon from the atmospheric system, you nudge the oxygen content up. Indeed, it increased enough to yield two competing hypothesis—oxygen surplus, and oxygen poisoning. In the former, the concept is supported by the evidence in the fossil record. Insects, who, lacking lungs, depend on passive air exchange across tiny pores in their abdomens to perform air exchange and bring in oxygen, seemed to have grown gigantic in this time. Dragonflies the size of seagulls?!?! Yep, that seemed to have happened to dragonflies and to other flying insect species. Living large in a world with more oxygen than ever seen before, these insects took advantage and became top predators in the air. With no avian competitors and plenty of energy from abundant oxygen, insects ruled the skies. As with any hypothesis gauging environmental conditions 300 million years ago, however, there is a competitor. The idea goes that there was so much oxygen that it was actually poisoning to insect life, so younger offspring were at an advantage if they had a larger body mass and thus a lower oxygen intake with respect to body mass as their more smaller framed competitors (https://www.nationalgeographic.com/science/article/110808-ancientinsects-bugs-giants-oxygen-animals-science#:~:text=According%20to%20previous%20theories%20about,Levels%20Fuel%20Mammal%20Evolution%3F%22). Whether insects were taking advantage of higher oxygen levels or were changing body size to adapt to it, it must have been a sight to see. Instead of the melodius song of birds flitting among trees, winged insects dominated the air, with the only sounds coming from their chitinous wings.

References

Algeo T, Scheckler S. Terrestrial-marine teleconnections in the Devonian links between the evolution of land plants, weathering processes, and marine anoxic events. Philosophical Transactions of the Royal Society B 1998;353(1365). https://doi.org/10.1098/rstb.1998.0195.

Algeo T, Scheckler S. Land plant evolution and weathering rate changes in the Devonian. Journal of Earth Science 2010;21:75–8. https://doi.org/10.1007/s12583-010-0173-2.

Algeo TJ, Scheckler SE, Maynard JB. 12. Effects of the middle to late Devonian spread of vascular land plants on weathering regimes, marine biotas, and global climate. In: Gensel PG, Edwards D, editors. Plants Invade the Land. New York Chichester, West Sussex: Columbia University Press; 2001. p. 213–36.

Becker R, Konigshof P, Brett C. Devonian climate, sea level and evolutionary events: An introduction. Geological Society, London, Special Publications 2016;423:1–10. https://doi.org/10.1144/SP423.15.

Berry C. How the first trees grew so tall with hollow cores – new research. 2017. https://theconversation.com/how-the-first-trees-grew-so-tall-with-hollow-cores-new-research-86150.

Bruckner M. The Gulf of Mexico Dead zone. 2017. https://serc.carleton.edu/microbelife/topics/deadzone/index.html.

Fields B, Melott A, Ellis J, Ertel A, Fry B, Lieberman B, Liu Z, Miller J, Thomas B. Supernova triggers for end-Devonian extinctions. Proceedings of the National Academy of Sciences 2020;117(35). https://doi.org/10.1073/pnas.2013774117.

Filippelli G. The global phosphorus cycle. Reviews in Mineralogy and Geochemistry 2002; 48(1):391–425. https://doi.org/10.2138/rmg.2002.48.10.

Filippelli G. The global phosphorus cycle: Past, present and future. Elements 2008;4(2): 89–95. https://pubs.geoscienceworld.org/msa/elements/article-abstract/4/2/89/137768/The-Global-Phosphorus-Cycle-Past-Present-and?redirectedFrom=fulltext.

Filippelli G. Phosphorus and the gust of fresh air. Nature 2010;467:1052–3. https://www.nature.com/articles/4671052a.

Filippelli G, Souch C. Effects of climate and landscape development on the terrestrial phosphorus cycle. Geology 1999;27:171–4.

Kenrick P, Strullu-Derrien C. The origin and early evolution of roots. Plant Physiology 2014;166:570–80.

Lenton T, Dahl T, Daines S, Mills B, Ozaki K, Saltzman M, Porada P. Earliest land plants created modern levels of atmospheric oxygen. Proceedings of the National Academy of Sciences 2016;113(35):9704–9. https://doi.org/10.1073/pnas.1604787113.

Marshall J. Prehistoric climate change damaged the ozone layer and led to a mass extinction. 2020. https://theconversation.com/prehistoric-climate-change-damaged-the-ozone-layer-and-led-to-a-mass-extinction-139519.

McGhee G. The late Devonian mass extinction: The Frasnian-Fammenian crisis. Columbia University Press; 1996.

Morris JL, Leake JR, Stein WE, Berry CM, Marshall JEA, Wellman CH, Milton JA, Hillier S, Mannolini F, Quirk J, Beerling DJ. Investigating Devonian trees as geo-engineers of past climates: linking palaeosols to palaeobotany and experimental geobiology. Palaeontology 2015;58:787–801.

Piombino A. The heavy links between geological events and vascular plants evolution: A brief outline. International Journal of Evolutionary Biology 2016. https://doi.org/10.1155/2016/9264357.

Smart M, Filippelli G, Gilhooly III W, Whiteside J. Decoding Devonian Mass extinctions: New evidence linking land plant expansion to marine anoxia. In: AGU Annual Meeting, San Francisco. American Geophysical Union, Annual meeting. American Geophysical Union; 2019. https://ui.adsabs.harvard.edu/abs/2019AGUFMPP54A.07S/abstract.

Smart M, Filippelli GM, Gilhooly III WP, Marshall JEA, Whiteside JH. Phosphorus in lacustrine sequences as records oof enhanced terrestrial nutrient release during the Devonian emergence and expansion of forest. GSA Bulletin 2022. in press.

Stein W, Berry C, Morris J, Wellman C, Beerling D, Leake J. Mid-devonian Archaeopteris roots signal revolutionary change in earliest fossil forests. Current Biology 2020;30(3): 421–31. https://doi.org/10.1016/j.cub.2019.11.067.

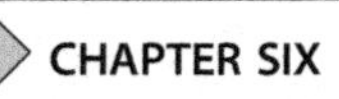

CHAPTER SIX

Massive extinction drivers and climate impacts

Introduction

Each of the Big Five Mass Extinctions decimated most of the life on the planet (hence the term "mass extinction"), and the Sixth one seems headed that way. Mass Extinctions are a bit hard to define, and the nature and impacts become harder to interpret the further back in time we go. But they are all caused by climate change. Whether that climate change occurs as a slow but steady venting of greenhouse gasses from the Earth's interior during a major volcanic interval, or from the opposite—a "nuclear winter" of freezing conditions following a major asteroid impact on the planet—a change in climate has caused massive shifts in Earth's major ecosystems. These shifts occurred when most life was found in the oceans, such as during the Ordovician Period, or appearing as the finale for the Age of the Dinosaurs just 66 million years ago. Mass extinctions provide a critical window into the boundaries of many Earth systems, and may also provide insights into how our own Sixth Mass Extinction might play out, and how ecosystems might respond, adapt, or perish in the future. Natural analogs are imperfect at providing guidance for our own future, as this is one case where the power to create, and likely to moderate if not completely stop, our own Mass Extinction lies with one organism—humans. But one reassuring observation we can make about all previous Mass Extinctions is that the Earth, and life on it, persisted. Much changed, of course, but still a planet with life and robust feedback cycles that kick in to stabilize climate again. https://www.amnh.org/shelf-life/six-extinctions.

The term Mass Extinction is variously defined as "... a widespread and rapid decrease in the biodiversity on Earth. Such an event is identified by a sharp change in the diversity and abundance of multicellular organisms. It occurs when the rate of extinction increases with respect to the rate of speciation" (https://en.wikipedia.org/wiki/_event Extinction), and "The extinction of a large number of species within a relatively short period of geological time, thought to be due to factors such as a catastrophic global

Climate Change and Life
ISBN: 978-0-12-822568-4
https://doi.org/10.1016/B978-0-12-822568-4.00001-8

event or widespread environmental change that occurs too rapidly for most species to adapt" (https://www.dictionary.com/browse/mass-extinction). Most comprehensive profiles of Earth's history provide a blow-by-blow recounting of the science of each of the six mass extinctions but suffice it to say that many common features are observed. This chapter is instead a narrative of mass extinctions that goes beyond a recounting of dates and deaths and data and focuses instead on several key aspects of earth systems that are illustrative of Earth resilience. Granted, our understanding of the drivers and impacts of these extinction events is constantly evolving, especially for the earlier events with poorer representation in the geologic record. Even the most recent one that killed off the dinosaurs 66 million years ago has only recently revealed many of its secrets. Additionally, many questions hover around our current, human-driven mass extinction, because the full climatic trajectory, and other human activities, which is driving it are up to our own future choices.

The big six—a summary

The drastic appearance of free oxygen on the planet likely drove the first major extinction after the Great Oxidation Event, but this event about two billion years ago and the relatively modest global ecosystem at that time together have not risen to the level of classification as a Mass Extinction. Instead, you need to fast forward Earth's history by about 1.5 billion years to the Ordovician Period about 445 million years ago. At this time, not much in the way of land plants or animals existed, or at least we have few extant records of them. But the marine record reveals a decline of about 85% of the Ordovician species Fig. 6.1.

The best current guess for a driver of this event is that rapid and repeated sea level oscillations intermittently exposed and then submerged shallow coastal habits. Because the coastal environment is nutrient-rich and full of light, being shallow, this is where the major ecosystems, such as reefs, thrive. Similarly, coastal environments on land may also contain complex ecosystems. But neither of these ecosystems can survive intact when the sea level varies wildly, and instead, just those opportunist species remained under these environmental pressures. The cause for the sea level variations is not well known, but may be related to major tectonic events at this time and thus major shifts in ocean basin capacities—when basin sizes decrease, but the volume of ocean water remains the same, sea levels rise and coastal regions are inundated. The reverse is also the case.

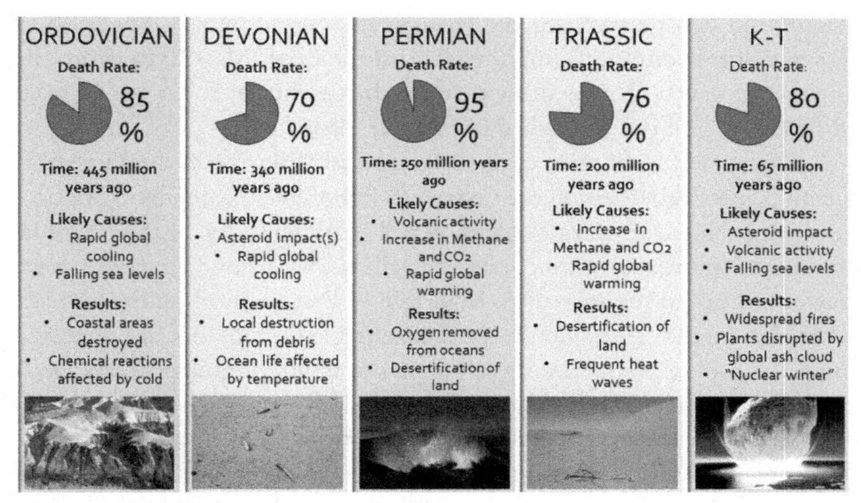

Figure 6.1 ***The Five Big Mass Extinctions*** **Mass extinctions involve significant and rapid declines in biological diversity and have had a host of contributing factors through time.**

The late Devonian Mass Extinctions events about 370 million years ago were covered in Chapter 5, but briefly, the evolution and proliferation of land plants with deep roots transformed the global ecosystem through an increased supply of limiting nutrients to the ocean. Massive eutrophication ensued, with perhaps some assistance from the nutrient dissolution of large volcanic eruptions at this time. Together, the enhanced burial of carbon in ocean sediments caused a reduction in carbon in the atmospheric reservoir and thus global cooling. Even though we have good geological records of this event, it is still difficult to separate multiple potential driving forces, as some of them "push" the Earth's climate in the same way—i.e., both deeper soils and volcanic eruptions increase ocean nutrient supplies—and result in the same net effect—eutrophication and ecosystem die-off.

The Permian-Triassic extinction 252 million years ago is likely also the product of volcanoes and extreme eutrophication and will be dissected later in this chapter, along with the Cretaceous-Tertiary mass extinction that killed off the nonbird dinosaurs. Another mass extinction occurred at the boundary between the Triassic and Jurassic periods, during the time of the Age of the Dinosaurs, about 201 million years ago. This one, marked by the demise of about one-third of marine species and many of the reptiles that existed at this time (making ecological room for the classic Tyrannosaurus Rex, apatosauruses, and velociraptors of movie fame), is the subject of significant debate, with climate change, an asteroid impact, or a spate of massive volcanic eruptions implicated. And the Sixth Mass Extinction, the human one, gets a chapter of its own, with a particular focus on the near and far future and the ecological fate of our planet.

At the heart of all of these extinctions is a basic true-ism of this incredibly resilient planet. Whether it is the first evolution of free oxygen by cyanobacteria, a global ice-over during the Snowball Earth episodes, or the multiple, current, and future mass extinctions, they all revolve around climate change. They also speak to the incredible resilience of Earth—through fire and ice, nothing has been able to quench the amazing spark of life that first appeared from the complex chemical soup of our planet about four billion years ago, a fact that gives universal hope that many other such beautiful blue, green, and white balls are spinning around their own stars scattered about the cosmos. When a planet has all the right ingredients, like the Earth does, and the range of potential changes to its systems is moderate, like for the Earth, the resilience of life reigns. For this reason, the first step to understanding the biological impacts of climate changes, and the potential recovery or replacement of ecosystems on the planet, relies on an understanding of the principles of resilience itself.

Resilience, and failure, of biological systems

Mass extinctions are defined as intervals of significant biodiversity loss and ecological transformation. But what is the tipping point for biological systems? How do ecosystems resist, and eventually succumb to, environmental change? How far can systems be pushed before minor to moderate disruptions become a mass extinction event? It all comes down to resilience and an understanding that many possible "ecological plateaus" can define what occurs when a system is pressured due to changes in internal or external forces. At a broad level, resilience is easily understandable, but it helps to look at some of the ways scientists have modeled resilience. For instance, for ecosystems, simple mathematical and graphical representations show how systems can find ways to exist based on their environment, the resources within it, and the changes that come to it. Consider, for instance, a pond with brook trout in it. The trout like cold water, and they like water that is not too acidic. If the pond becomes too acidic (as happened in many northern ponds due to acid rain) the trout are driven to refuges that are less acidic, or they stop reproducing. Other fish may survive better, as the pond becomes a new, though possibly less productive, ecosystem.

Resilience models are one simple way that many scientists represent such a change Fig. 6.2. The hills and valleys represent the conditions affecting the ecosystem, with the ball representing the current state of the system. At higher levels, the ball represents a more productive and ecologically healthy system. If stress comes along (acidic water), the ball can be pushed from valley 1 (where the trout are thriving) to valley 2 (where a new acid-tolerant species has come along). The ecosystem survives but is in a less productive state. To move back "up the hill" toward ecosystem health requires a change in the conditions that pushed it downward in the first place, and typically requires more "energy" or resources than it took to drop the ball down the hill in the first place. The overall system has two states that it tends toward, one desirable state and one undesirable state. The resilience of the desirable state (the upper valley) is linked to the factors that determine how difficult it is for the ball to roll out or roll back up the hill in the event of a downhill state change. Most ecosystems are of course more complicated than this, as some have many different drivers that push one way or another, much more so than in a small pond. Nevertheless, systems tend to have various stable states, and a move "down the hill" to a less desirable state can be difficult to reverse.

In this resilience model, the unifying idea is that systems tend to find a balance somewhere. This idea of *homeostasis* is vital to the analysis of physical

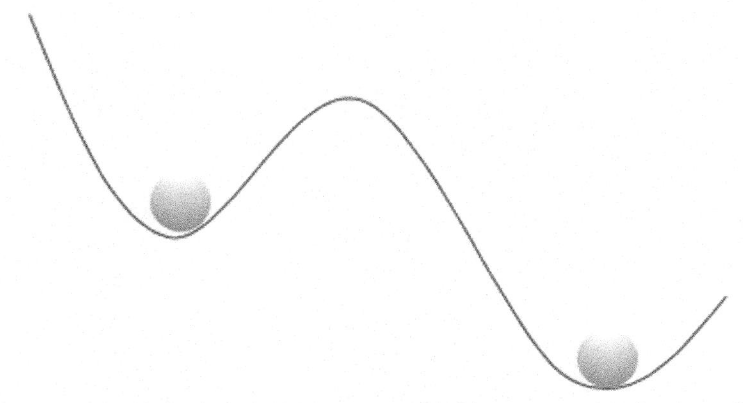

Figure 6.2 ***Ball and String Model of Resilience*** Stable environmental states of fish in ponds can be achieved under a number of conditions. At higher levels, the ball represents a more productive and ecologically healthy system. If stress comes along (acidic water), the ball can be pushed from valley 1 (where the trout are thriving) to valley 2 (where a new acid-tolerant species has come along). The ecosystem survives but is in a less productive state.

ecosystems, species adaptation, and ultimately human happiness and health. Homeostasis simply means a state in which things tend to persist, much as in the human body when all organs and functions work together to produce the stable state that we know simply as "life." When something threatens that state, either the organism will adapt to a new state to survive (resilience) or it will die.

Given this, it is clear that organisms and ecosystems that have equilibrated and thrived in one set of climatic conditions can become unstable if the climate changes quickly, or changes to extremes outside of its resilience window. Indeed, the term "resilience" is a critical one in considering any system. Take for example a coral reef. The base of a coral reef ecosystem is, of course, corals themselves Fig. 6.3. These organisms are fascinating, and inhabit a relatively large range of environmental conditions, from tropical shallow margins to the deep sea. But perhaps the most iconic reef is the Great Barrier Reef on the eastern Australian continental margin. This reef is not one monolithic thing, but rather a 2286 km long (equates to 1429 miles long and 133,000 square miles in areal extent) system of reef structures with a diverse subset of habitats and reef compositions. It has been around for at least 600,000 years, and thus constitutes one of the oldest and most stable ecosystems on the planet. But even this variety and longevity are being challenged by human-induced climate change, and the very future of this system will depend on the resilience of the ecosystem to the changes already impacting it.

Figure 6.3 ***Healthy Coral Reef*** The healthy coral ecosystem at Norman Reef.

The greatest threat to coral reef organisms is surface ocean heating. Ocean heat waves reveal the fascinating lifestyle of corals and the resilience that these organisms can display in the face of external pressures. To get a full understanding of this resilience, it is important to understand a bit about coral biology. Thousands of subspecies of corals exist, with thousands of individual morphologies and colors, and growth habits. But one commonality is that coral is a half animal-half plant hybrid. Not too far separate in fact from all other organisms on the planet, who evolved billions of years ago to incorporate foreign material into their cells, which now serve critical biological functions. Human beings, and all other animals, have "foreign" DNA in our cells in the form of mitochondria, which serve as the energy engines that fuel cellular function but also have a different DNA than the rest of our bodies. Interestingly, this mitochondrial DNA is only inherited from your mother as it is based solely on the DNA in eggs. In the case of coral, it is easiest to think of the host cell as the "animal" part, which is called a polyp.

Polyps have the ability to extract calcium and carbonate from the water and make hard parts—the coral skeleton. These structures act much as our own skeleton does—to provide verticality and three-dimensional structure to the coral colony for stability in the tidal environment where corals tend to grow and for space to feed. The polyps themselves inhabit this structure and feed by filter feeding, a process where they extrude sticky tendrils into the water and pull in food in the form of organic material that falls from the surface or is swept along by currents. But coral polyps long ago adopted a version of the mitochondrial approach to succeeding as an organism, by incorporating plant cells into their tissues. These plant cells are called dinoflagellates, and they are very common in marine environments. The coral polyps recruit these cells from seawater to be part of an amazing symbiotic relationship—once recruited, they are termed "zooxanthellae" ("yellow animal" in Greek, because the dinoflagellates incorporated are typically yellow in pigment). The coral structure provides safety from predation and the coral polyps provide excreted nutrients from respiration. The dinoflagellate cells in turn use these concentrated nutrients, plus incoming sunlight, to produce sugars, which they "share" with the coral polyps as a supplementary and stable food source. The zooxanthellae are moved to the surface of the polyp during the daytime to achieve maximum photosynthesis potential and pulled back at night when the filter-feeding components of the polyp are typically extruded.

In conditions within a certain metabolic window, this relationship between coral polyp and zooxanthellae is a relatively stable one. But even before humans came on the scene, there is evidence that the relationship has its own boundaries. Coral polyps can, and have, ejected their plant roommates in times of stress. These events are called coral bleaching events, because, upon the ejection of their plant symbionts, the corals lose their coloration and turn white Fig. 6.4. The colorful coral forest that you see snorkeling or diving is largely the palette of the zooxanthellae themselves, which alter the reflected color from the surface by absorbing light in various wavelengths according to their type of chlorophyll. Lose the plant cells, and you lose the color.

A bleached reef does not always spell ultimate doom for a reef system. The coral polyps can "re-recruit" their dinoflagellate roommates should environmental conditions stabilize, surviving meanwhile in their bleached state by filter feeding alone. But this time window of survivability is short—typically on the order of a few years. If better conditions do not return, then that coral head, or reef system, dies. To be accurate, the bulbous and branching structure of the coral skeletons themselves can persist, eerily like a field of skeletons, and can become the home of noncoralline organisms like algae. But once a reef fails, it is very difficult for coral organisms to recolonize the system.

This window of conditions, and time, that characterize coral resilience is being challenged now, as marine heat waves become more frequent, and more persistent, in many tropical areas. In 2022 alone, roughly 90% of

Figure 6.4 ***Bleached Branching Coral*** Branching coral that has been bleached by a marine heat wave. Bleached corals can recover if the environment soon returns to a suitable one for their algal symbionts.

the Australian Great Barrier Reef experienced bleaching. Likely much if not all of that reef area will fully recover, but many marine biologists see these recurring ocean heat wave events, which are strongly linked to anthropogenic climate change, to be the ultimate survival challenge for tropical coral reefs around the world. And in the case of coral systems, it is not just the heat that causes the symbiotic dinoflagellates to "raise a red flag" and be emitted by coral polyps, but the increased acidity that comes with higher carbon dioxide as well. Carbon dioxide in the atmosphere dissolves into surface seawater, a process much like the carbonation of your soda, which is dissolved carbon dioxide that is injected into your favorite beverage upon bottling.

Much like your soda, a portion of the carbon dioxide that dissolves into seawater converts to carbonic acid. For your soda, that is a good thing, as the acidity balances out the otherwise too-sweet soda (ever taste a warm flat soda–terrible, right?). For seawater, the carbonic acid lowers the pH of the ocean, making it more acidic. Acid is the bane of the mineral calcium carbonate, which is the building block of many marine organisms, including corals. Humans have increased the concentration of carbon dioxide in the atmosphere by 50% in a short 150 years, which in turn means that the acidity of the ocean has increased by about the same amount. In the short term, the. increased acidity makes it more metabolically difficult for calcium carbonate-dependent organisms like corals to grow and thrive. This increased "ocean acidification" is thus another ecological stressor to add to marine heat waves when considering the resilience of coral reef ecosystems in a human-dominated world. In the longer, geologic term, the excess acidity will be buffered by ample stores of marine sediments that are comprised of calcium carbonate minerals, but the timeframe of this natural buffering action is on the order of 10,000 years. Coral reefs as we know them are not likely to dominate tropical areas even 80 years from now due to the dueling pressures of heat and acidity.

So why dwell on coral biology, coral reef resilience, and threats to coral reefs in a discussion about mass extinctions? Well, it is exactly this type of massive ecological die-off that is captured in the geological record, and by which we mark past mass extinction events in the first place. Because of the ample skeletal remains, broad spatial coverage, and unique structures that will be left should tropical reefs ultimately fail, it will be easy for a geologist several million years in the future to proclaim that the coral demise signals a mass extinction event. But by understanding the processes that are driving a potential mass extinction event occurring before our very eyes, we can better understand the extreme variations in climate change impact life, and what extremes might push an ecosystem over the edge to extinction. Granted, and as discussed elsewhere here, the rate of change that humans are exerting on the planet's climate systems is probably only exceeded by the extreme events that led to the demise of the dinosaurs, and thus is perhaps only a great analogy for this event and not for the other, slower acting mass extinction events in the geologic past.

Vicious volcanoes

Volcanoes seem to be the bane of organismal existence on this planet. Not just volcanic eruptions burying ecosystems (and towns) under ash and lava, but the climatic influence that they have on longer time scales. Of the five past Mass Extinctions, it has been argued that most if not all were driven or aided by volcanic processes. Take for example the worst Mass Extinction, the Permian-Triassic extinction 252 million years ago. The evidence that this is massive is written clearly in the geologic record—70% of land species and 80% of marine species that were thriving at the end of the Permian were permanently erased at the Permian-Triassic Boundary, apparently victims of extreme climate change (https://www.physicsforums.com/threads/todays-climate-change-and-the-permian-triassic-boundary.995178/www.physicsforums.com/threads/todays-climate-change-and-the-permian-triassic-boundary.995178/). Permian rocks yield huge layers of coal, which were formed in swampy terrestrial environments full of trees and plants and which still hold substantial fossil remains of these organisms. Then, just above these black-ribboned deposits sit buff-colored sandstones with nary a fossil to be found. This contrast—vibrant dense forests giving way to sterile sandy layers, corresponding to a shift of geochemical compositions, which indicate a rapid increase in the concentration of carbon dioxide in the atmosphere, along with resultant extreme warming. The mechanisms for the carbon dioxide release may be the massive volcanic eruptions at this time, deposits of which are called the Siberian Traps. These deposits contain three million cubic kilometers of volcanic rock—consider that volcanic eruption from the ever-active volcanoes on the big island of Hawaii amount to less than 10% of that volume (https://pubs.geoscienceworld.org/gsa/geosphere/article/9/5/1348/132675/Modeling-volcano-growth-on-the-Island-of-Hawaii0) to appreciate the size of the Siberian Trap eruptive event. The intrusion and eruption of the Siberian Trap into preexisting overlying Permian rocks heated the coal, oil, and gas that formed from these swampy deposits and released so much carbon dioxide that the concentration increased to six times previous levels. Charcoal from sedimentary layers can clearly be seen in overlying layers, the result of heating and burning of the organic-rich layers from volcanic intrusion. The sharp build-up of greenhouse gasses in the atmosphere that ensued caused a global heatwave, with estimates of a temperature rise of 10−14°C lasting over tens of thousands of years.

But this direct impact of volcanism on climate change is not the end of the story. Warming and high levels of acidic carbon dioxide caused an extreme and sustained state of continental weathering, particularly from the freshly erupted volcanic rocks, releasing massive amounts of nutrients, particularly phosphorus, from rocks into rivers, lakes, and ultimately the ocean (https://www.nature.com/articles/s41467-021-25711-3www.nature.com/articles/s41467-021-25711-3). We know exactly what ensues when all of these nutrients end up in waterways—algal slime dominate. Algae blooms that contaminated Lake Erie and shut off the water intake for the entire city of Toledo in 2015 (https://theconversation.com/climate-change-threatens-drinking-water-quality-across-the-great-lakes-131883) and cause the vast Dead Zone in the Gulf of Mexico resulted from an excess of nutrients, but have catastrophic ecological impacts. They choke out native ecosystems, and upon dying, the consumption of the bloom material by zooplankton sucks up all of the oxygen in the surrounding water, killing any animals around. Evidence of such a massive and sustained eutrophication event exists in early Triassic rocks, where plenty of bacterial and algal remains are found but few animal fossils (https://www.scientificamerican.com/article/toxic-slime-contributed-to-earth-rsquo-s-worst-mass-extinction-mdash-and-it-rsquo-s-making-a-comeback/). Earth went from a planet with complex terrestrial ecosystems in the Permian to one choked out by slime in the earliest Triassic (https://www.nature.com/articles/s41467-018-07934-z).

Collisions, climate change and mass extinctions

Before launching into the dramatic events leading to the extinction of the dinosaurs, it is important to step back and consider galactic dynamics. As discussed earlier, climate change is a product of both external and internal forces acting on the planet. External forces include the amount of solar radiation hitting the planet and the seasonal timing of that solar radiation. Both of these factors are dictated by the very predictable and periodic orbital variations that Earth experiences—the Milankovitch Cyclicity. Internal forces include the balance of carbon between the Earth's deep reservoir and its surface reservoir, as well as the net reflectivity of the planet. Mass extinctions occur when the climate shifts significantly in one direction—going from cold to hot, for example. And for most mass extinctions in the Earth's record, scientists have looked largely at these internal forcing functions. Intervals of significantly enhanced volcanic activity, for example, can tip the carbon balance such that a significantly higher fraction of carbon is in the surface reservoir, including the atmosphere, resulting in a sustained interval of higher temperatures and consequent ecosystem disruptions. But one theory for mass extinctions that has lingered over the decades has nothing to do with these internal mechanisms, and instead is aligned with the type of catastrophic asteroid collisions that spelled the demise of the dinosaurs.

Interstellar theory

We think of our Solar System as comprised of a Sun, eight planets (poor Pluto ...), a remnant shattered planet between Mars and Jupiter (the Asteroid Belt), and nothing else. But in fact, we have a vast swarm of debris that formed or accumulated at the outer edges of our solar system that occasionally terrorizes the planet. The famous Halley's Comet spends most of its time there, as do the other major comets and a whole host of asteroids. This debris-cluttered region at the outskirts of our solar system is called the Oort Cloud. Many stray remnants of early solar system formation spend their time permanently in the distant Oort Cloud, but some, like the dramatic comets that swing by the Sun periodically, have more ovoid orbits. The orbit of Halley's Comet, for example, has a period of about 75 years, and last entered the inner solar system in 1986 Fig. 6.5.

Many of the larger asteroids have similar types of orbits, although they have not been mapped with the same regularity as comets because the distinctive nature of these tailed wonders fascinated human civilizations for millennia, and was recorded with enough detail as to be added to our current astronomical reckoning of comets. Asteroids and comets have struck the Earth in the past and will do so in the future, simply by the blind misfortune that involves two bodies orbiting the sun in different orbital shapes but over billions of years—eventually, even an unguided billiard ball will hit another one. Some of the larger extraterrestrial objects have been called "civilization ending" bodies because if they were to strike the Earth now, they would cause a series of physical events that would lead to extreme disruption, if not a total annihilation of all that we know. Recent estimates by NASA of such an event occurring in any given year are quite low (something like 0.000,001% annual probability- https://interestingengineering.com/what-is-the-probability-of-a-huge-civilization-ending-asteroid-impact), but we have seen how disruptive even a smallish impact could be to our planet. An asteroid that was only 50–60 m in diameter exploded over the Earth with a force of 12 megatons in 1908. It struck in nearly completely uninhabited Siberia but leveled about 80 million trees over an area of 2150 square kilometers of forest Fig. 6.6.

The so-called Tunguska Event is more of a scientific curiosity thanks to its remote location but had it exploded over New York City or London in 1908, the devastation would have been enormous. Because of the potential existential risk posed by these extraterrestrial bodies, an entire program of space observation, NASA's Near Earth Object program, works every day to identify, catalog, and track these planet killers. Even with this system in place, surprises do occur—for example, a asteroid 100 m in diameter passed between the Earth and the Moon in 2019 that was both uncharted and large enough to wipe out a city. Another such event was dramatically recorded by the Hubble Space Telescope in 2014—this time, a string of massive cometary fragments slammed into Jupiter and caused significant disruption (see "Did you know that?"). And occasional interstellar interlopers whiz through our solar system, not from the Oort Cloud but from deep space (https://www.space.com/oumuamua.html). As any fan of apocalyptic science fiction will tell you, there seem to be many right, and wrong, ways to deflect an incoming asteroid from hitting the Earth point blank and averting the end of days. But for now, not only are we stepping up our ability to forecast such an event, the likelihood of "death by asteroid" remains remarkably low in the span of a human lifetime, but certainly not in the span of geologic time.

One hypothesis that has been lingering around the fringes of earth sciences for several decades is that the apparently periodic nature of extinction events is tied to periodic intervals of high Earth bombardment by extraterrestrial material. The most vocal proponents of this hypothesis, Michael Rampino and Ken Caldeira, argue that the Solar system passes through the mid-plane of our galaxy every 26 million years in its rotation around the galactic hub (https://www.theatlantic.com/science/archive/2015/11/the-next-mass-extinction/413884/). This region of space is very dust-filled, and the hypothesis goes that the mass of materials disrupts the orbits of comets and asteroids in the Oort Cloud and sends them flying through the inner solar system, increasing the odds of an Earth impact.

Figure 6.5 ***Halley's Comet*** **Halley's comet last passed through the middle of the Solar System in 1986 and will return again in 2061.**

The solar system is actually in this danger zone right now and will remain so for the next several million years. Before searching for real estate in a different solar system, however, other earth scientists have called this hypothesis into question, arguing that both the periodicity of the extinctions and the definition of "extinction" proposed for the 26 million-year cycle are poorly defined. Nevertheless, we certainly can point to one major collision 66 million years ago as a very disruptive global event—one that would qualify as a civilization-ending event if it occurred today.

Dissection of a dino-killer

A suspicious half-ring of cenotes, or open sinkholes, around the northern Yucatan Peninsula of Mexico were the first local clue that something strange had happened here. Together with geologic evidence scattered around the planet from about 66 million years ago, a dividing line between the Cretaceous and Tertiary Periods, between the Age of the Dinosairs and the Age of the Mammals. These disparate pieces of evidence have slowly but inevitably pulled together to provide a snapshot of a very bad day about 66 million years ago. This journey of discovery starts in central Italy in the hills near Umbria, then moves to the deep Atlantic Ocean off the Blake Plateau and the plains of North Dakota, before being captured north of the Yucatan

Figure 6.6 ***Trees Leveled by the Air Blast from the 1908 Tunguska Impact Event in Siberia*** The Tunguska Impact released a significant amount of energy and resulted in a shock wave that leveled thousands of square kilometers of trees. Fortunately, it occurred in an extremely sparsely populated area in Siberia.

Peninsula in perhaps the most unequivocal pieces of geologic evidence ever recovered from the deep past. The "perhaps" in the preceding sentences is because even this slam dunk of an event has another competing, or at least additional, hypothesis to explain it, one that makes the Cretaceous-Tertiary Boundary mass extinction follow in line with the other mass extinction events in the geological record.

Alvarez and iridium

The father and son team of Luis and Walter Alvarez were geologists working at the Lawrence Berkeley National Laboratories Fig. 6.7. Together with colleagues, they noted in 1980 that the rock layers that formed at the very distinctive boundary between the Cretaceous and Tertiary Periods were extremely rich in the extraterrestrial element iridium, found in high amounts in asteroids but not on the surface of the Earth. The huge enrichment of this layer with iridium, at levels up to 160 times the normal level. The geologists found the clearest evidence for the excess iridium in a clay layer in the hills near the picturesque central Italian town of Gubbio, which also yielded a definitive date coinciding with the Cretaceous-Tertiary boundary. This finding in Italy was corroborating by iridium-rich layers in sedimentary

Figure 6.7 ***Luis and Walter Alvarez*** The father-son team of Luis and Walter Alvarez suggested that the high iridium values widely found in a layer of rocks at the Cretaceous-Tertiary boundary (here near Gubbio, Italy) represented the impact of a large extraterrestrial body at that time, likely contributing to the demise of the dinosaurs.

layers across the globe and led Alvarez and colleagues in 1980 (Science) to propose a massive extraterrestrial impact coincident with the well-documented extinction of the nonavian dinosaurs at this same time. It was logical to link this cataclysmic impact with an equally cataclysmic impact on Earth's biota, including its dinosaurs. The only thing lacking in 1980 was a "where," of the asteroid impact, but given that 70% of the Earth's surface is covered by ocean, and that the ocean crust that might hold a crater remnant of this impact is frequently recycled through plate tectonics, it could simply have been erased from the geologic record.

The "Alvarez Hypothesis" was formally adopted as a leading candidate for the dinosaur dye-off during a conference convened in Snowbird Utah in 1981 and inspired geologists on a global hunt to find the location of the impact. Unbeknownst to this group, the crater location was being presented at nearly this same time in a small meeting by geophysicists Glen Penfield and Antonio Carmago after oil-exploration research funded by Pemex in the northern Yucatan. Their geophysical observations revealed magnetic and gravity patterns indicating a massive ring-shaped crater in this area. The geophysical surveys and earlier cores were taken for oil exploration indicated an impact severe enough to melt deep rocks and crack the Earth's crust. But the geophysicists faced internal criticism of their interpretations by Pemex, and it went no further than this after the geophysicists, after reading the Alvarez paper in Science in 1980 and writing to the team, received no response.

The global fallout

Over the following two decades, more and more evidence for a catastrophic event mounted. This evidence included vast areas of tektite accumulation and shocked quartz at the time of the dinosaur extinction. Whereas the iridium-rich layer was a geochemical indicator of an extraterrestrial impact, the tektites and shocked quartz were direct evidence of high temperatures and thermal and seismic shock from the impact. Tektites are small droplets that form when an asteroid hits the Earth, melting rocks at the impact site and ejecting them from the force of the impact into the atmosphere as a vast cloud of glassy spherules, which rain back down to Earth within minutes to hours of an impact event. Shocked quartz is exactly what it sounds like—normal quartz minerals, which are typically quite resistant to physical alteration, have their mineral structures "shivered" by the impact blast. Between iridium, tektites, and shocked quartz, the evidence was mounting to support the smoking gun, an asteroid, but still no gun to be found, until the 1990s. Finally, people were talking about the hypothesis, the earlier geophysical evidence from Yucatan was resurfacing, and new satellite imagery revealing the bottom half of a wide ring of cenotes in the northern Yucatan was published—collectively, these conversations began solidifying this location as the leading contender for the impact site. The nearby northern Yucatan town of Chicxulub became the namesake for the crater and the impactor.

Subsequent work has revealed many fascinating dynamics of the syn- and postimpact world at that time. For example, deep-sea sediment core samples taken by the Ocean Drilling Program (the predecessor of the current International Ocean Discovery Program– https://www.iodp.org/) off of the Blake Plateau revealed a "death layer" in the ocean, marked by a thick ash layer, no fossils, and a sharp change in ocean ecosystems before and after the event. It also provided clear age dates and was an incredible validation of both the timing and the global impacts of the event. Careful drilling operations yielded a beautiful section that shows the entire boundary and the marine dynamics surrounding that boundary Fig. 6.8.

Before the event, this area experienced normal marine biological sediments, including microfossils of large relatively complex zooplankton that thrived in the time of the dinosaurs. Then, the moment of impact caused a disruption in the marine sediments and was marked by a layer of glassy nodules (tektites) from the impact ejecta. This layer of atmospheric fallout may have constituted days to months of time, and the thickness reveals the quantity of material that was deposited at this site. The location of the drilling site is down-blast from the Chicxulub crater and is in the Atlantic Ocean, on the other side of the Florida peninsula. Thus, it might have more glassy fallout debris than other marine sites farther away. Nevertheless, what became obvious in this core was that nearly all of the complex zooplankton that existed before the impact is no longer present, presumably driven to extinction by severe disruption of the marine food web. Just above the ejecta layer is one dominated by soot and other fallout from the global fires that were sparked by the tremendous amount of heat energy resulting from the impact, and likely sparked by the fallout of hot ejecta. Above the fireball layer is normal marine biological sediments, but of completely different composition than before the impact, being dominated by simple microfossils that survived the impact. This one core, now replicated and present in several museum sites constitutes a clear view of the global havoc wreaked by this meteor impact.

Other sites show the full array of ecosystem impacts, one of which is particularly impressive. A fossil site in North Dakota included what researchers called a "postimpact snapshot" of events from thousands of kilometers north of Chicxulub (https://www.pnas.org/doi/10.1073/pnas.1817407116). This snapshot was marked by tektites found in preserved amber, the tell-tale enriched iridium layer, and mass death. About 50% of the fossilized fish had tektites embedded in their gills—they might literally have suffocated from them. These fossils were embedded in a dense sedimentary layer that appears to have hardened like concrete after being rapidly deposited on these hapless organisms from a tsunami-like water movement that was triggered by the magnitude 10–11 earthquake generated by the impact. As a reference, the largest, most destructive earthquakes that have struck in recorded history are all well below magnitude 10, and this scale is logarithmic meaning that a magnitude 10 quake is 31 times stronger

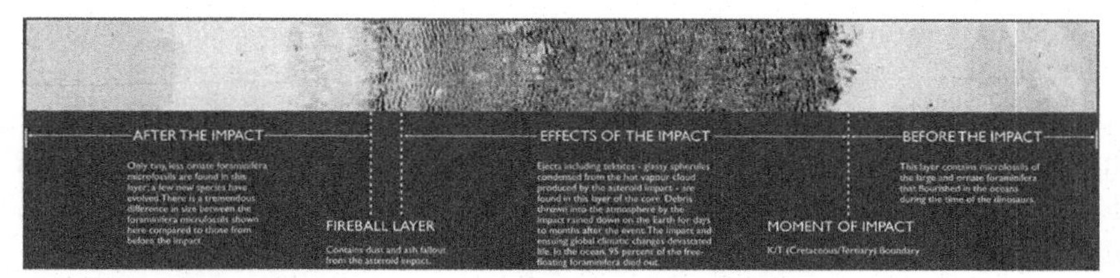

Figure 6.8 ***A Core from the Blake Nose Showing the Cretaceous-Tertiary Boundary Sequence*** A deep-sea sediment core sample taken by the Ocean Drilling Program (the predecessor of the current International Ocean Discovery Program– https://www.iodp.org/) off of the Blake Plateau, revealed a "death layer" in the ocean, marked by a thick ash layer, no fossils, and a sharp change in ocean ecosystems before and after the event.

than a magnitude 9 quake (https://earthquake.usgs.gov/education/calculator.php). And it wasn't just the direct seismic impact that is reflected in the fossil record here, but also the impact-related tsunami—one fish appears to have been broken in half after being flung into a tree from the force of the tsunami. A 2022 study went so far as to suggest that the Chicxulub impact occurred during the Northern Hemisphere spring based on geochemical records from these fossilized fish. So now, not only do we know that it was a bad day for the dinosaurs 66 million years ago, but it was a bad day in April!

Other monstrous asteroid impacts have left their mark on the planet, and many of them seem not to be implicated in driving major ecological disturbances as Chicxulub did. One of the most iconic and visible craters is Meteor Crater, sometimes called Barringer Crater, in Arizona, a National Landmark that has fascinated visitors for decades Fig. 6.9.

Figure 6.9 ***Meteor Crater, Arizona*** Meteor Crater, near Flagstaff Arizona, formed from an impact of a 50 m wide asteroid about 50,000 years ago.

It is indeed an extremely impressive site as it is young-ish (~50,000 years), is contemporaneous with modern humans on the planet, although not likely in this area, and is well preserved by the desert location. The meteor that created this wonder in the desert, however, was minuscule as these impactors go. It has been estimated at 50 m across (half a football field), big enough to blast a big crater into the Earth's surface but only 0.5% the size of the Chicxulub asteroid. To be sure, we have evidence of Chicxulub-type impactors in the Earth's records, but you have to go a lot farther back than 50,000 years before present to see them. For example, the Sudbury Basin in Canada is the geological remnant of a Chicxulub-size impactor that struck the planet 1.8 billion years ago, as is the Vredefort crater in South Africa from 2 billion years ago. Although there was life on earth at the time that both of these struck, it was very simple single-celled organisms, and no biological evidence remains of any disturbances coinciding with the impactors. This is not to say that these events were benign, as they have left a ~2-billion-yearold scar on the planet's surface, but the paleontological record is simply too sparse at this time to provide evidence supporting coeval ecological disturbances.

Why did the Chicxulub impact have such an outsized impact on the ecosystems of the planet, given its relatively small size? At 10 km across, it is the obviously large by our standards, being the size of many medium-sized cities in the United States, for example. But compared to the diameter of the Earth, it is puny, at only 0.08% the size. This is comparable to one red sand grain on a plate with 1270 white sand grains—you would likely not even notice it among the sea of white. But the key difference is that the meteor struck the planet going 20,000 km per second, or almost 45 million miles per hour! Although not a perfect analogy, imagine flinging that red sand grain onto the plate of white sand at 45 million miles per hour—the shock wave from one tiny grain at that speed would make a mess of the plate, to be sure. And indeed, the Chicxulub impactor made a mess of the Earth's crust. It blasted a 180-kilometer diameter crater, plowing 20 km into the Earth. This depth is near the boundary between the solid, cool outer crust of the planet and the semiliquid, hot, mantle underneath. This impact released the same energy as 100,000,000 megatonnes of TNT (4.2×10^{23} J), over a billion times the energy of the atomic bombs dropped on Japan at the end of WWII. Although analogies like this to big explosions are frequently used, it is likely that few reading these numbers were (a) in WWII, and (b) can conceive of what a billion times the force of the atomic bombs means. Perhaps most useful is again to compare the 50-meter Meteor

Crater impactor, which made a little divot on the desert landscape 50,000 years ago and might have killed some hapless rabbits on the landscape, versus the 10 km Chicxulub impact that caused a crust-shattering impact on the planet and destroyed the entire dominant and iconic lifeform at that time.

Location, location, location

But it was where it hit that seemed to wreak so much planetary havoc. It had a 70% chance of striking the ocean given how much of the Earth's surface is blue. An ocean strike would have caused a massive amount of steam and a tsunami of historic dimensions, and woe to the Mosasaur who might have been swimming along minding its own business before being landed on by a monstrous rock from outer space. But its impact would have gone largely unfelt by the land-dwelling dinosaurs at the time and would be mostly nonexistent in the geologic record. If it hit the other 30% of the Earth—the terrestrial surface—it would have similarly plowed a deep trench into the crust and thrown up debris from the now vaporized rocks that it struck into the upper atmosphere. This would have constituted a much greater planetary threat, given that the ejected debris in the atmosphere would have briefly cooled the planet, as modern volcanic eruptions do, and the rain of still blistering debris that would quickly land back on the planet would have ignited forests and singed hapless dinosaurs that were "down-blast" of the impact. But again, it is possible that this would have posed more of a hiccup than a heart attack for the planetary ecosystems. No tsunami would have resulted, and it could be argued that the marine environment would not be catastrophically impacted, and the dinosaurs might have managed to weather the storm as it were and persisted postimpact. Based on a long string of geologic detective work, we now know that neither an ocean nor land collision occurred, but Chicxulub instead, by chance, hit a critical zone between land and ocean. This proved to be a case of planetary bad luck, at least for the dinosaurs, but before we delve into the why of this let's explore the detective story.

Chicxulub, revealed

After decades of searching and research, and the ultimate identification of the impact site at Chicxulub, we had a good sense of the size, scale, and global magnitude of the event. But we had little data to ground-truth our guesses about the geological and geochemical characteristics of the impact site itself. That is until large scientific expeditions were mounted in the

2010s to uncover the many questions that remained. And by far the most definitive evidence for Chicxulub impact effects has come from the impact site itself, the culmination of a decade of exhaustive and persistent efforts to conduct deep sampling expeditions from ground zero itself to determine the size of impact, the crustal deformation that ensued, the chemistry of rocks at the impact site, and postimpact geologic and biotic recovery that occurred there.

After significant international negotiations and based on the amazing promises that direct samples from the impact site might hold, the International Ocean Discovery Program conducted Expedition 364 in 2016 in partnership with another major scientific exploration, program, the International Continental Scientific Drilling Program (https://www.icdp-online.org/home/). It was a challenging operation in that it occurred in only 20 m of water—far too shallow for normal ocean drilling shops to conduct operations. The scientists instead employed a type of jack-up rig Fig. 6.10. that served as a drilling platform and towed it to the location to begin operations Fig. 6.11. The drilling target was designed to determine the geology

Figure 6.10 ***Drilling Platform for IODP Expedition 364*** This "jack-up rig" allowed for scientific ocean coring in the shallow northern Yucatan Peninsula margin for IODP Expedition 364.

of the rocks where the impactor hit and how they were altered, as well as the postimpact recovery of the region.

Some of the important discoveries from this scientific expedition included identifying changes in the fabric of the Earth's crust and the post-impact recovery of the site. The geological tumult caused by the impact was documented by the fact that the crater ring left over from the impact is largely granitic crust from the deeper earth, uplifted by 10 km and frozen in place like the rim of Meteor Crater (but much more massive). Additionally, the crust was so fractured that vigorous circulation of fluids and seawater occurred through the cracks, setting up a strong hydrothermal system that persisted up to 100,000 after the impact before being "sealed up" by settling and mineralization of the crustal cracks through which fluids circulated. Finally, the recovered core material revealed that the crater site became a bit of an oasis for the recovery of ecosystems in the area, setting the stage for the biological transition into the post-Cretaceous world (https://phys.org/news/2020-11-placement-earth-largest-mass-extinction.htmlhttps://phys.org/news/2020-01-microbial-mayhem-chicxulub-crater.html)

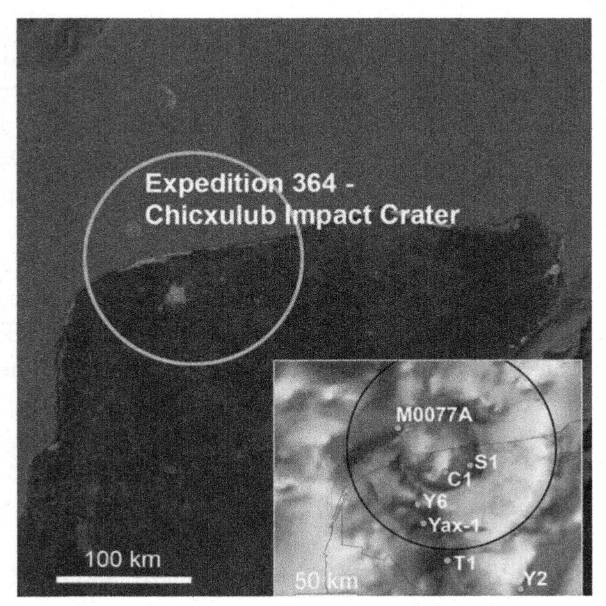

Figure 6.11 ***Location of Chicxulub Impact Crater, Yucatan, Mexico*** The IODP sites where the first core samples were taken of the peak ring of the Chicxulub impact crater. The inset diagram shows a geophysical reconstruction of the larger crater body.

(https://gizmodo.com/extreme-life-thrived-in-hot-asteroid-pit-after-dinosaur-1845532447). Indeed, recovery was so quick, in a geologic sense at least, that by 30,000 years postimpact the dynamic ecosystems that thrived in this setting were in full swing. The rapid recovery at the direct impact site might seem counterintuitive, but the vigorous hydrothermal circulation likely fed these systems with substantial amounts of rock-derived nutrients, thus supercharging this emerging oasis.

Minutes to millennia after the impact

As documented by coring and other examinations of the site, the Chicxulub impactor struck on the continental margin of what is now the Yucatan Peninsula (https://www.pnas.org/doi/10.1073/pnas.1909479116). In doing so, it struck thick layers of sulfur-rich carbonate rocks and evaporite deposits that were deposited in the shallow seas of this margin during the Cretaceous. Unlike granites or other more heat-resistant rocks derived from igneous processes, carbonate sediments vaporize at a relatively low temperature. In fact, most carbonate rocks break down at 900°C, as opposed to continental crust rocks like granite that don't vaporize until 3000°C. The heat at the heart of the Chicxulub impact site was in excess of 10,000°, well above these rock stability plateaus, but that was likely achieved in only a small area. The heat-susceptible carbonate and evaporite rocks in the general vicinity of the impact would have vaporized these materials and flung sulfur particulates far into the stratosphere. These particulates are known to reflect sunlight back into space before it hits the planet, and the aerial spreading of sulfur particles has been proposed as one way to counter human-produced global warming. The details and philosophy of this approach are troubling and will be explored in the final chapter, but suffice it to say that we know from past major volcanic eruptions that sulfur particles do cool the planet for several years after atmospheric ejection. With the amount that may have been produced by the vaporization of evaporite deposits from the collision site, these cooling particles may have lingered for a decade.

In addition to the sulfur particles, organic matter in the strike site rocks (https://www.pnas.org/doi/10.1073/pnas.2004596117) and soot from wildfires would have also exacerbated the cooling effect. Collectively, they sent the planet into what is called a "nuclear winter," which is a Cold War-era conception of what would occur on Earth in the aftermath of a mutually destructive nuclear war. Sunlight would be severely reduced, stifling plants and primary production for years and basically cutting off the base of the food chain. And, of course, global temperatures would drop. Based on geochemical evidence, this temperature drops soon after the Chicxulub impact was extreme, plummeting the tropics to 5°C from the preimpact balmy temperature of 27°C (https://www.nature.com/articles/srep28427). The extreme temperature drop plus the lack of food would have killed off terrestrial and marine ecosystems. But if that weren't enough, immediately after the nuclear winter sustained warming caused by all of the carbon dioxide released from dead biomass raised global temperature back up to about 5°C higher than before the impact, a condition that persisted for roughly 100,000 years (https://www.science.org/doi/10.1126/science.aap8525). The thermal whiplash would have been extreme, likely far exceeding the resilience windows of many ecosystems at this time.

A volcanic trigger?

The preceding has focused on one trigger—a large impactor—and the subsequent impacts. If this alone was the cause of this fifth mass extinction, then it would be quite distinctive from the other mostly volcanic-related extinctions. But a set of geoscientists scientists, including Gerta Keller from Yale and Paul Renne from the University of California, Berkeley, are not ready to give all of the credit to meteors, arguing that a major volcanic eruptive sequence at the same time as the dinosaur demise certainly cannot be a coincidence. The well-dated Deccan Traps in India are massive volcanic flow deposits that formed about 66 million years ago and would have resulted in the emission of a significant amount of warming carbon dioxide into the atmosphere Fig. 6.12. (https://www.nytimes.com/2015/10/06/science/study-finds-asteroid-ahead-of-dinosaur-extinction-accelerated-volcanoes.html?_r=0.). This is consistent with the measured increase in global temperatures that occurred soon after the chilling nuclear winter—to sustain a warming event for 100,000 years likely requires more carbon dioxide than could be added even by combusting much of the Earth's biomass.

Support for the potential role that the Deccan Trap eruption played in climatic upheaval after the impact comes from a large set of samples collected from the deposit. The Deccan Traps are huge, with the basalt formation covering more than 200,000 square miles and reaching more than a mile thick in some places. Geochemical analysis of these samples revealed that the Deccan eruptions, which may have begun before the Chicxulub impact, doubled in output within 50,000 years of the impact.

Figure 6.12 ***Deccan Traps*** Thick volcanic layers from the Deccan Traps in India are evidence of substantial increases in volcanic eruption rates around the time of the Cretaceous-Tertiary Boundaries, 66 million years ago.

These researchers argue that this doubling of eruptions on the other side of the planet after Chicxulub impact was no coincidence. The impact had enough force to literally crack the tectonic plate where it stuck and send seismic shock waves around the world. The argument goes that this seismic shock wave encircled the globe and effectively cracked wider the volcanic conduit that was already bringing molten lava from the deeper Earth to the surface in India. Other than model reconstructions and the timing of the Deccan eruption increases, there is no further current direct evidence for the link, but like many Earth processes, it is not inconceivable that both a meteor impact and an impact-induced increase in volcanic activity both contributed to the substantial climate shifts that resulted in the end-Cretaceous Mass Extinction.

Summary

Much can be learned about Earth's coupled biogeochemical systems by examining Mass Extinctions. Much like the resilience that systems showed during the first evolution of free oxygen around 2.1 billion years ago and then again with multiple Snowball Earth intervals 1.5 billion years later, life found a way through multiple Mass Extinctions that have altered the course of life on the planet. With all of these intervals and events, the role that climate change plays in biological and ecological trajectories is profound. Indeed, we owe that ascendancy of our own group, the mammals, to the most recent mass extinction at the Cretaceous-Tertiary boundary, which tilted the organismal scales toward mammals, a tilt that has been reinforced over the past 66 million years as the Earth has shifted toward more mammal-friendly Ice House conditions.

The long dominance of reptiles and dinosaurs on the planet from 252 to 66 million years ago was perhaps largely due to a relatively uninterrupted time of very warm and stable climates, with consistently high levels of carbon dioxide in the atmosphere and plate tectonic conditions amenable to climatic stability. In contrast, the rapid and persistent climate swings that define the latter part of the Cenozoic Era have helped to forge rapid diversification of mammals that have uniquely evolved for colder and more variable conditions. The Age of Humans is markedly different from the Age of the Dinosaurs, not just because of colder conditions but also due to human ability to rapidly alter the planet. Our era-ending catastrophe won't likely be a massive meteor, but instead the intensity of resource exploitation, including the combustion of fossil fuels and rapid climatic changes that ensue, upon which modern society has been built and which will continue to define our species for the foreseeable future. The lessons of planetary resilience will still hold true, albeit pushed to the limits due to the Chicxulub-like rapid global change that is occurring from human activities. It is humbling to consider that the remains of our seemingly permanent civilization will likely similarly be available for scientific exploration 66 million years in the future. And if this chapter provides insight into anything, it is that rest assured—there will be life on the planet 66 million years into the future. It won't be humans as we know it, but it will be spectacularly interesting!

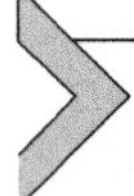

Did you know that?

We uncovered the potentially devastating effects that a meteor impact would have on the planet during Cold War experiments with above-ground nuclear testing

For over a decade, one main activity of the nuclear programs of the two nuclear superpowers (the USA and the former Soviet Union) was building and exploding ever-larger and more sophisticated nuclear weapons. Much of this testing was done above ground in the early days, and it was during this testing that the concepts of air blasts or shock waves emanating from massive above-ground explosions became clear. This allowed scientists to begin modeling the airwaves from past (and potentially threatening future) meteor impacts. From this work, it has been estimated that the air blast from the Chicxulub impact would have generated winds in excess of 1000 km per hour, an unimaginable force that would have leveled trees, and dinosaurs, on a continental scale.

The Tunguska Event revealed the scope and scale of what even a relatively small meteor would do when it struck, but we have direct observations of the devastation that a tiny meteor can cause, even one that broke up and vaporized in the atmosphere. In 2013, 105 years after Tunguska, another meteor caused problems in this Russia. The Chelyabinsk Event was widely observed, with many videos taken by observers, on the morning of February 15, 2013, in the southern Ural region of Russia Fig. 6.13. The meteor was only 20 m in diameter, and hit the earth at a very shallow angle, causing the object to explode and vaporize completely within the atmosphere without striking the surface. The atmospheric shock wave from this explosion was equivalent to 400—500 kilotons of TNT, or 26—33 times as much energy as was released by the detonation of the atomic bomb over Hiroshima. About 1500 people were injured during this event, and over 7000 buildings were damaged. Many of the injuries came from imploding glass windows, but some suffered temporary light blindness and even sunburn from the brightness of the explosion. A shocking aspect of this event is that the meteor went undetected until it entered the atmosphere, eluding our network of telescopes meant to warn of these events.

Figure 6.13 ***Photograph of the Light Streak from the Chelyabinsk Meteor*** The Chelyabinsk Event was widely observed, with many videos taken by observers, on the morning of February 15, 2013, in the southern Ural region of Russia.

The impact of Shoemaker-Levy into Jupiter was almost missed by the hubble space telescope due to computer glitches

Astronomers Carolyn Shoemaker, Eugene Shoemaker, and David Levy discovered a comet in March 1993 that looked unusually stretched out. It was eventually determined that the comet had been ripped apart by Jupiter's gravity into numerous fragments, and they were on a direct collision course with Jupiter from July 16 through 22, 1994. Having never directly observed two extraterrestrial solar system bodies collide before, no one knew exactly what they could expect to see when this happened. Astronomers and space enthusiasts across the world were determined to capture this unprecedented event.

This was a perfect opportunity for the Hubble Space Telescope, especially after it had its telescopic "contact lens" installed during a space servicing mission the year before to correct aberrations in its mirror. During this mission, it also had upgrades to its computer and other hardware and was lifted up a bit in its orbit after some orbital decay. The Hubble was primed and ready to capture the light show that Shoemaker-Levy was lightly to put on in July. Then, on July 4, 1994, the Hubble spontaneously put itself into a safe mode, after some type of self-diagnosed malfunction. A classic case of an overreacting computer and terrible timing coinciding. Mission engineers went into full panic triage mode, as they only had 11 days to get Hubble working again before the big event. They soon tracked to issue to a memory unit that had failed on the new flight computer. They realized that luckily there was a backup memory unit available from an earlier Hubble servicing mission available for use, but it was not yet configured for the new computer. Normally, reconfiguration would have been a longer, more cautious process, as mistakes made could cripple a telescope operating remotely 340 miles above the surface of the Earth. But the clock was truly ticking, and bad luck struck again—as they rushed to reconfigure the memory, Hubble spontaneously put itself into an even deeper safe mode.

This time, the onboard computer thought that two of the six new gyroscopes malfunctioning. They focused on determining whether the gyroscopes, which are critical for pointing Hubble to where it thinks it is pointing, were truly malfunctioning as the new onboard computer was indicating. This seemed unlikely, as it is not normal for multiple gyroscopes to fail at the same time. Eventually, NASA engineers tracked the issue back to a gyroscope counter protocol that was not reset when the new computer system was installed. This was then quickly resolved, and Hubble was back up and running in the nick of time, capturing one of the most incredible sequences of images of the comet string Fig. 6.14. and the massive atmospheric waves that appeared in the clouds of Jupiter afterward Fig. 6.15.

One intriguing and logical potential driver of extinction events has never panned out

Many organisms rely on the Earth's magnetic field to gauge which direction is north and which is south, using internal compasses. Some do this for migration purposes, especially birds who need to follow the seasons for food and breeding grounds. They employ magnetic-sensitive particles for this sense of direction (https://www.frontiersin.org/articles/10.3389/fphys.2021.667000/full). Others employ the magnetic field to find their

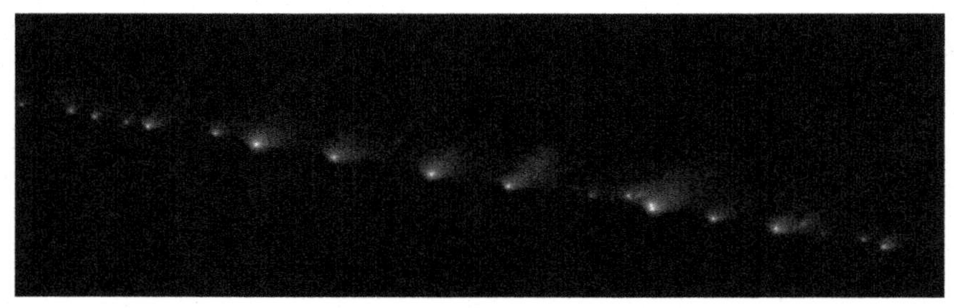

Figure 6.14 ***Strong of broken pieces of the Shoemaker-Levy comet. The Shoemaker-Levy comet string is observed by the Hubble Space Telescope as it approaches Jupiter.***

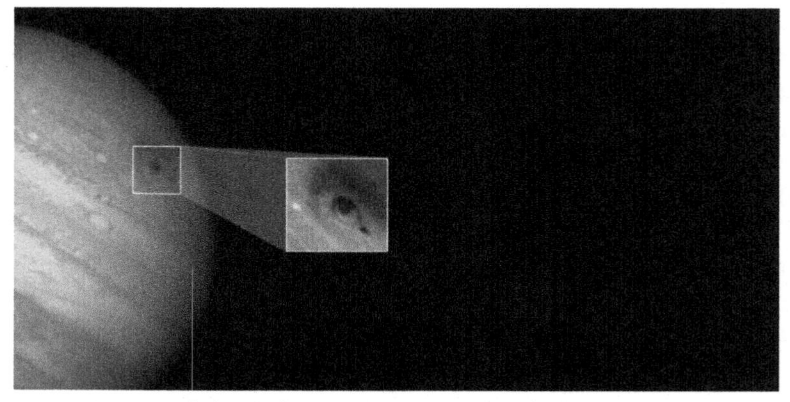

Figure 6.15 ***Impact of Shoemaker-Levy Comet into the Jovian Clouds*** The collision of the string of Shoemaker-Levy comet debris made a lasting mark on the clouds of Jupiter.

birth area, including sea turtles who lay their eggs on the same beaches where they themselves were hatched. Sharks use this sense to detect prey. Even termites use it—the huge magnetic termite mounds of Australia are built with a fluted structure that aligns with the Sun to minimize direct sunlight and thus stay cool in the blazing Australian summer Fig. 6.16.

Geoscientists have wondered why this amazing ability doesn't become a substantial liability when the Earth's magnetic field weakens, and the magnetic pole wanders, and occasionally flips completely north to south. Shouldn't these geomagnetic reversals so confuse these magnetic-sensing organisms that birds migrate the wrong way, sea turtles die of exhaustion trying to find their home beach, and magnetic termites build their mounds in the exact wrong way, cooking instead of cooling them? Many studies have sought a link between magnetic reversals and species extinction, but none has been confirmed. Granted, the field reversals occur sporadically, and

Figure 6.16 ***Magnetic Termite Mound*** Magnetic termite mound in the Northern Territory, Australia. These mounds are positioned in relationship to the sun and are designed to keep termites at a stable temperature, even during the Australian summer.

the last one was 781,000 years ago, but it seems that there should be some commensurate bird die-off or other extinction event caused by magnetic disorientation. Perhaps these magnetic organisms use more than one line of evidence to know the right direction, including the Sun angle as a function of season, which doesn't change with magnetic orientation. We clearly don't give other animals enough credit!

CHAPTER SEVEN

From greenhouse to ice-house: the coevolution of life and climate through the Cenozoic

Introduction

It started with a planet-devastating fiery impact, resulting in the fifth Mass Extinction that the planet has ever experienced. And it ended, 66 million years later, with the evolution of an organism having the capabilities, and short-sightedness, to be causing the planet's sixth Mass Extinction. Welcome to the Cenozoic Era! Although the Cenozoic has pretty catastrophic bookends to it, this chapter of the Age of the Mammals is an amazing one. We have an adequate geologic record to have made significant headway into measuring and at times even understanding, the processes that have linked climate change and life together. From interior Earth processes and volcanism to continental plates drifting and oceans closing, and opening, and entire ecosystems evolving in response to changing climates, the Cenozoic brings a reasonably clear record of earth system processes. It also paints a picture of a resilient planet, but one whose climatic swings are extreme, showing no preferential treatment for any organism or ecosystem. Unlike the long and consistently warm Era before it when the dinosaurs reigned supreme for hundreds of millions of years, the Cenozoic is marked by a slow but consistent march toward colder and colder conditions.

Try as we might, the human combustion of fossil fuels will never take us back to dinosaur times, when atmospheric carbon dioxide concentrations were in the thousands of parts per million. Throughout the Cenozoic, most of that carbon made its way back out of the ocean and atmosphere environment, locked up in geologic materials. It was in no way a steady monotonic decline, but the net result of that shift of carbon to the geologic reservoir was cooling temperatures. Additionally, changes in the configurations of continental and ocean basins caused significant "zonality" in climate, such that the polar regions are markedly colder than the tropics, a situation that has not always prevailed in Earth's history. Essentially, heat does not move from the equator to the poles very easily in this current tectonic

Climate Change and Life
ISBN: 978-0-12-822568-4
https://doi.org/10.1016/B978-0-12-822568-4.00007-9

configuration, leaving both poles to be uniquely isolated from the global energy flux arriving from the Sun. The transition from Greenhouse to Icehouse conditions is one driven by geological processes and amplified by the role that greenhouse gases play in modulating global climate.

Scores of books could be (and have been) written about the Cenozoic climate, but in this chapter, we will explore this history in illustrative vignettes rather than a blow-by-blow description. This includes (1) what drove a still mysterious extreme but temporary warming event that is moderately analogous to what we are currently causing and experiencing with anthropogenic climate change, (2) the role that the drifting continents have played on global heat transfer and climate, (3) how changing biogeochemical cycles influenced the evolution and diversification of the organism whose shells now filter your beer, (4) what happened when a large chunk of Antarctica broke off and plowed into Asia, (5) the silica arms race between horses and grasses, and (6) what we know about how the Earth responded during a recent interval of fairly high atmospheric carbon dioxide. Some of these processes are truly finite events, whereas others occurred over millions, or tens of millions, of years.

As a brief summary, some of the longer-term tectonic events during the Cenozoic included the continued opening of the Atlantic Ocean Basin and the continued closing of the Pacific Ocean Basin. Although the Pacific remains the largest of the ocean basins, it is ringed with very active margins on all sides, the so-called "Ring of Fire" where oceanic lithosphere and plenty of island systems are being recycled into the deep earth. Thus, the margins have plenty of earthquakes and volcanoes marking the sometimes violent nature of plate recycling. The Cenozoic also saw the separation of Antarctica from its connection with South America via the opening of the Drake Passage, an event that caused an entirely new pattern of ocean circulation that is responsible for the distribution of so many modern marine ecosystems, and weather systems. During this period, a circum-equatorial ocean circulation pattern ended for good following the closure of seaways both in Europe, with the collision of Africa into Europe, and in America, with the uplift of Panama and the closing of the Isthmus of Panama. This latter event occurred only about 4 million years ago and paved the way not only for the isolation of the Caribbean Sea but also for the need to dig the Panama Canal back through this narrow isthmus for expediting shipping traffic. These major marine tectonic events collectively reshaped how heat is distributed on our planet, and helped to define the trajectory of the evolution of marine and terrestrial ecosystems through the Cenozoic.

From an ecosystems perspective, the Cenozoic saw the various machinations of mammalian evolution, but also various shifts in major land plants and marine organisms. Horses and all of the large megafauna (think elephants and giraffes and blue whales) evolved throughout the Cenozoic, slowly replacing many of the ecological niches occupied by the dinosaurs. Grasses evolved, and began to proliferate toward the latter part of the Cenozoic. And this is not just the grass that you have to mow in the summer, but the full gamut of a plant type called a C4 plant (corn, for example) that is uniquely suited to make the most of the lower carbon dioxide concentrations that persisted in the later Cenozoic. The marine phytoplankton was dominated by types that secrete shells based on silica, such as diatoms, proliferating in a number of ocean ecosystems such as the equatorial and polar regions and displacing carbonate phytoplankton.

Extreme warming, PETM style

Whether the Age of the Dinosaurs ended with a bang, or perhaps an extended whimper followed by a bang, as proponents of the volcano-driven extinction would argue, is still being debated, but one thing is definite—not a single nonavian dinosaur group Fig. 7.1 survived the Chicxulub impact 66 million years ago. In fact, three-quarters of all Earth's biota died out in the

Figure 7.1 ***A Casuari*** These Australian birds hearken back to the group of avian dinosaurs that survived the Chicxulub impact and persisted, and flourished, in the post-dinosaur world.

thousand or so years after the impact event. Whether you consider the lumbering Tyrannosaurus Rex on land or the massive Mesosaurus in the sea, you start the Cenozoic with a veritable "blank slate" upon which new organisms evolved, and new ecosystems formed. By all evidence, they did that reasonably quickly (https://www.washingtonpost.com/news/speaking-of-science/wp/2016/11/07/how-long-did-it-take-for-life-to-rebound-after-the-death-of-the-dinosaurs/), with some estimates or ecosystems returning to preimpact biodiversity after a geologically short 4 million years in South America, and 9 million years in North America (https://www.nature.com/articles/s41559-016-0012). Some of these estimates come from how much insects were chomping on plant leaves found in the fossil record—as insect numbers go, so goes ecosystem health and diversity. (Sadly, as discussed in the final chapter, the Earth has already lost nearly 50% of insect diversity due to climate change and chemical-intensive farming practices, which certainly will show up when geologists 66 million years from now uncover our human strata). The difference between the rates of recovery on the two continents might be due to the fact that North America was effectively at ground zero of the Chicxulub impactor, or at least directly "downstrike" from it so that landscape devastation was more profound than on any other continent.

The post-Chicxulub recovery in terms of biodiversity might have been complete after 4–9 million years, but organisms on land, and in the sea, looked completely different than before. First, the main larger animal groups were mammals and birds. Both of these groups have much higher resting metabolisms than do reptiles or the nonavian dinosaurs before them because they have to burn calories to maintain a constant internal temperature. They are calorie hogs with a need for consistent food sources, but they also were more consistently active as they didn't depend on passive heating or cooling, and thus they were capable of continual food foraging.

In the sea, the large swimming dinosaurs were gone, but many fish groups remained and continued thriving after the impact. And a new group was soon to come into the marine scene—marine mammals, such as whales and dolphins, and manatees. Their evolutionary history is fascinating, as they originated as land-based organisms but descended into the depths where they slowly but surely lost most of the land-related physical characteristics, such as legs and tails, which had no purpose in the water. They didn't lose their need to breathe air, albeit they did evolve the capacity to hold their breath for extremely long periods of time, a trait necessary for deeper food foraging or even for napping in a predator-free submarine environment.

The interval after the Chicxulub impact was reasonably balmy, being about 10°C warmer than today and continuing the general temperature trends from the Age of the Dinosaurs. The polar areas were also much warmer relative to today as there was consistent heat transfer from the tropics to the poles. Of course, that doesn't mean they were always warm—the planet was still tilted after all, and the poles endured at least 3 months per year with no sunlight, with their only heating coming from the tropical transfer, but that was enough to ensure that no ice persisted through the summer. Seen from outer space, the Earth would have looked much different than today, with none of the distinctive white ice cover at the poles and on Greenland, even though the continents would have been more or less recognizable.

In many respects, the planet was just climatically chugging along, even after the impact disaster, with the main difference a change in actors, but not a change in set design. This 10 million year-long recovery and stabilization period (called the Paleocene Period), however, was soon to be shattered by a still-mysterious, hotly debated, yet profound shake-up to the Earth's climate system roughly 55 million years ago. Termed the "Paleocene-Eocene Thermal Maximum" (or PETM for short), because it marked the divide between the Paleocene and the Eocene Period which was to follow, and because it was devastatingly hot, the PETM might provide a window into what happens when earth systems experience a massive injection of carbon over a short amount of time. The PETM thus provides a useful, though imperfect, natural experiment that mimics some aspects of humans' own massive injection of carbon into earth systems. In the case of PETM, however, we can see how the systems recover, how long that recovery lasted, and what if any longer-term effects lingered.

The PETM has been the subject of an array of scientific efforts and expeditions, including some that have taken place via drilling and recovery of the sediment record by the International Ocean Discovery Program (IODP) (Fig. 7.2).

Several "IODP" expeditions have focused largely on recovering pristine marine records from this interval, with the advantage of often being of higher temporal resolution and less prone to disturbance than layers that are found exposed on rock outcrops on land. Of course, a limitation with IODP sampling is that it is restricted to marine settings, and thus terrestrial processes are a bit more indirectly reflected in the sediment cores that are

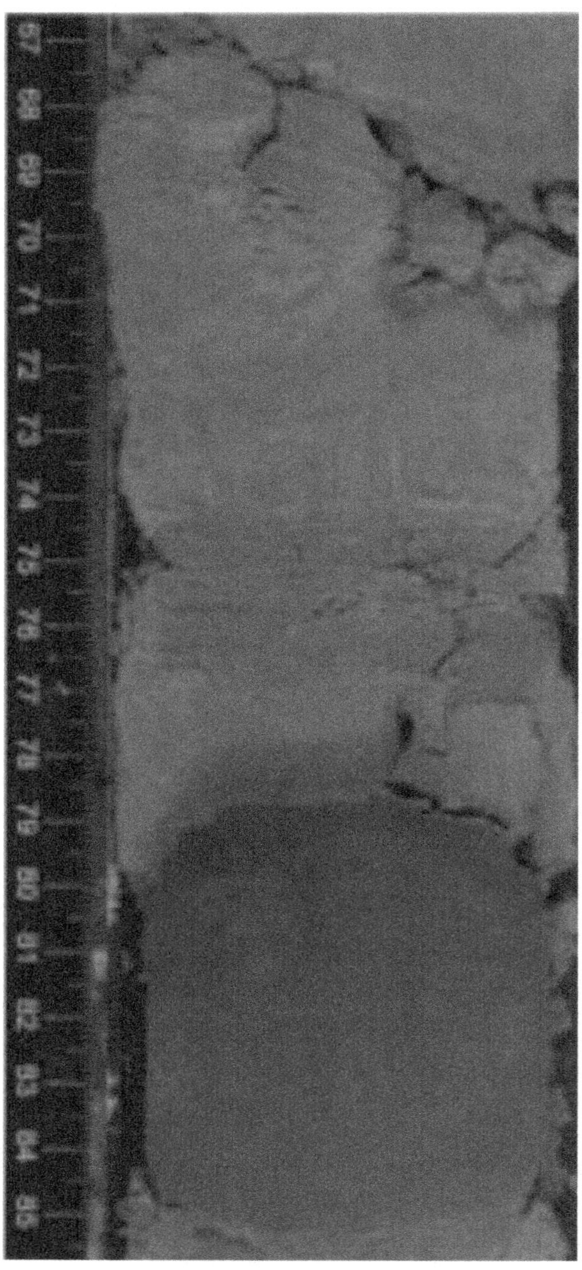

Figure 7.2 ***Core from IODP Expedition 342*** This core shows the initiation of the PETM, with the dark brown base representing a major ecological shift to silica-producing zooplankton at this time.

recovered. But the cores speak volumes about the profound changes that occurred during the PETM, including extreme temperature fluctuations, massive shifts in the global carbon cycle, marine organism extinctions, and a large ocean acidification event that impacted many of the carbonate-dependent marine organisms (think the equivalent of modern coral reefs).

The PETM began with an extreme increase in global temperatures, occurring within a few thousand years and seeing the ocean and land surface temperatures increase by more than 5 C (9°F). This global temperature increase is roughly on par with what we might expect over the next several centuries if we took no actions at all to reduce carbon emissions, leading to a. run-away greenhouse. Fortunately, many policy initiatives and market shifts paint a rosier picture of less extreme temperature increases within this century at least, but of course that gives no indication of what global temperatures might look like in several centuries. If we take the progression of the PETM as an imperfect analogy, what we would see is significant widespread extinctions from climate change alone, and particularly severe ocean ecosystem disruptions. The cause of our current sharp temperature increase and overall climate plight is human activities, injecting tremendous amounts of carbon into the atmosphere. The cause of the PETM event was likely a similarly massive and rapid release of carbon into the atmosphere. This release was obviously not triggered by the actions of sentient beings unless alien visitors were having a heyday tinkering with our geological record. A clear shift in the carbon isotopic composition of various fossil and organic deposits indicates that this carbon emission spike was from methane, which has a uniquely extreme isotopic signature.

The release of massive amounts of methane might have been driven by subseafloor volcanic eruptions (https://royalsocietypublishing.org/doi/abs/10.1098/rsta.2010.0053) or by changes in ocean circulation (https://agupubs.onlinelibrary.wiley.com/doi/full/10.1029/2001PA000678?casa_token=5RxN4cFVsCQAAAAA%3AxyjJR3dP3Aul0tTXv6vZdcjuWcSWfMsSNxQjWEowaKwfBz5fah0BzHEKwYJQAW6dwOsv5U7-52Y_kg), or both. It is likely that one or more of these events changed the stability regime of the vast storehouse of mineral-bound methane that is in a precarious balance with marine temperature and pressure conditions. This mineral methane is called methane hydrate or clathrate and is actually an ice-like form of methane that, like ice, is not particularly stable. Clathrates exist today in all ocean basins and many polar lakes, and climate scientists rightly keep an eye on their stability to ensure that current global warming is not tipping the balance toward massive destabilization of these clathrates and the release of the methane gas that they hold. Should this occur in a widespread way, it would be catastrophic as the released methane is an extremely powerful greenhouse gas, which would then warm the planet, which would then melt more clathrate, and so on. And this seems to be exactly what happened in the PETM, and the resultant vicious cycle of warming went on until effectively all the vulnerable clathrate was dissolved. The resultant methane "burp" warmed the planet, but it also oxidized to carbon dioxide, which continued to warm the planet, and also dissolved into the oceans, causing ocean acidification. This latter is evidenced by substantial declines in marine carbonate fossils, with the process of carbon dioxide dissolving into carbonic acid in the ocean resulting in the dissolution of shell material—organisms were fighting a losing battle as their shells literally dissolved away Fig. 7.3

(https://www.pmel.noaa.gov/co2/story/Contact+Us). Furthermore, the injection of carbon into the atmosphere enhanced the weathering of continental rocks (through carbon dissolution into rainwater causing carbonic acid), resulting in an increase in the supply of some nutrients to the ocean (https://pubs.geoscienceworld.org/gsa/geology/article/44/9/731/195232/Silicate-weathering-and-North-Atlantic-silica?casa_token=FopYZm5ZTvcAAAAA:vfvddmTtuLHgftP6y4RgRRiH3WgC3Q0CEUixFlOCcfFpzbpaVSWtrlreIQ-epTUYj_jSq68) that was eventually was offset by negative feedback loops in the carbon cycle.

Figure 7.3 ***Bivalves Shells Dissolving*** Increased acidity of seawater causes the shells of organisms that secrete calcium carbonate to literally dissolve away. *Courtesy of David Littschwager/National Geographic Society.*

There are limits to how much we can compare the PETM and its impacts over time with our own situation of anthropogenic climate change. First, the PETM occurred during a time that was already warm, and when there was little to no continentally-based ice to melt away. Thus, the extremely warm conditions that ensued after carbon injection didn't raise sea level from glacial or ice sheet melting, but only from thermal expansion. Even this effect would be significant with 5°C of warming—making an assumption that the entire water column warmed by this much (not a great assumption, mind you, but a starting point), this would come out to a PETM-induced sea level rise of about 16 m (~50 feet), which is a not insignificant change (https://www.scientificamerican.com/article/seas-may-rise-23-meters-per-degree/#:~:text=Environment-,Seas%20May%20Rise%202.3%20Meters%20per%20Degree%20C%20of%20Global,research%20institute%2C%20released%20on%20Monday). But this is nothing like the ~120 m sea level variations that occur during current glacial/interglacial cycles. Additionally, heat transport at this time was more "global" with less of a gradient between the equator and poles than exists today. From that, one can assume less of a sharp impact on polar ecosystems beyond the ambient global changes that the PETM caused. So overall, even this significantly disruptive event, one of the most rapid shifts in carbon cycling and global temperatures over the Cenozoic, is likely to pale in comparison to how a comparable, but human-induced, change in carbon cycling will impact the planet.

The PETM lasted for about 100,000 years, a number that we know quite well thanks to successful efforts from the IODP (e.g., http://publications.iodp.org/preliminary_report/342/index.html). By recent comparison, that is roughly one glacial/interglacial cycle in length and is roughly consistent with our understanding of the response time for the short and medium-term carbon cycles. In this way, it does appear to be a transient event that produced extreme variations in the carbon cycle that were accommodated for by the carbon sinks that exist on land (plants, trees, soil) and in the ocean (calcium carbonate sediments). Following the PETM, global temperatures dropped back to previous levels. Other than a shorter-term reversal to warmer temperatures (during the "ELMO" interval, short for Eocene Layer of Mysterious Warming and obviously a clever Sesame Street-inspired acronym in search of a name), gradual cooling set in and began accelerating at the end of the Eocene and the subsequent Oligocene Epochs, likely driven by plate tectonics and a continued decline in atmospheric greenhouse gas concentrations. The rest of the story of Earth's climate and life through the Cenozoic is one of the ever-cooling temperatures, and the ecosystem responds to a variety of environmental changes, some subtle and others spectacular.

The Earth's geodynamo ... and Mt. Everest!

Much of the story of Earth's longer-term climate variations are written in the mountains and on the seafloor. Extreme mountain uplift and subsequent erosion, driven by the collision of tectonic plates, not only leaves clear signatures in the mountain rocks and the sediments eroded from them but also substantially drives climate through the influence of weathering on the carbon cycle. Similarly, the speeding up and slowing down of Earth's internal dynamo are clearly seen in the magnetic stripes left on the seafloor. Because the planet's geodynamo flips directions sporadically, the magnetic minerals in seafloor basalts and sediments overlying them follow suit, as iron filings moved around by a magnet in science class. Frequent flipping of the geodynamo, which yields more closely-spaced magnetic stripes on the seafloor, indicates a very energetic tectonic state for the planet, with widely-spaced stripes indicating the opposite. A more energetic inner Earth means that it recycles its stored inner carbon more quickly to the surface, typically yielding higher atmospheric carbon dioxide concentrations and a warmer climate. Interestingly, through the middle and end of the Cenozoic, both seafloor spreading and mountain building may have acted to drive the carbon balance toward a new extreme. But both ideas continue to be the subject of vigorous geologic debates—the Earth doesn't give up its secrets that easily, it seems.

First, let's talk stripes ... on the seafloor. By conducting a global inventory of seafloor magnetic records, geomagnetists (or geomagicians, as they are referred to by insiders), have found various relationships between the speed of the geodynamo and its net impact on Earth systems. We can actually measure the rate of the current Atlantic basin opening with Global Positioning System (GPS) sensors on either side of the ocean. Current spreading rates are about 2–5 cm/yr (an inch or two) with some variation along the 16,000 km long mid-Atlantic Ridge (https://www.sciencedirect.com/topics/physics-and-astronomy/mid-ocean-ridges). The last time that our geodynamo fully flipped directions was 781,000 years ago at the Brunhes-Matayama Magnetic reversal (https://www.google.com/search?q=brunhes-matayama+&rlz=1C5CHFA_enUS825US825&sxsrf=ALiCzsYWmyjWDQGAwzM9oexezY0V8kjP5A%3A1651644208035&ei=M

BdyYuXkAeaNseMP-K-30AQ&ved=0ahUKEwjlvMKSlsX3AhXmRmwGHfjXDUoQ4dUDCA4&uact=5&oq=brunhes-matayama+&gs_lcp=Cgdnd3Mtd2l6EAMyBAgAEA0yBggAEA0QHjIGCAAQDRAeMgYIABA-NEB4yCAgAEAgQDRAeOgcIABBHELADSgQIQRgASgQIRhgAUN-gOWKpZYJNiaAJwAXgAgAGDA4gB7QmSAQcwLjIuMS4ymAEAoAEByAEIwAEB&sclient=gws-wiz), and thus the current magnetic direction seafloor stripe (with the north magnetic direction to the north) is roughly 16–39 km wide. Should it flip tomorrow, the new seafloor basalts and sediments would take on the reverse pattern, with north to the south.

One long-standing paradigm is that the rate of seafloor spreading has gone down substantially over the Cenozoic. Yes, the areas of the ocean that are growing, like the Atlantic Basin, are still doing so, but at a slower rate. A consequence of the "geodynamo slowdown model" is that carbon is not being recycled efficiently out of the Earth's interior, causing the observed decline in atmospheric carbon dioxide and global temperature through the Cenozoic. Another consequence, supported by coastal geological records, is that slower seafloor spreading rates mean that the oceanic lithosphere itself is cooler and thus less buoyant, sinking to a lower absolute level and creating additional ocean basin storage for seawater. This would mean that sea level, with respect to the geography of the continental lithosphere at least, would be lower, which indeed it is. Sadly, as is often the case with science, newer reconstructions dispute this slowdown, meaning that perhaps the process of tying a bow on this tidy story is not yet complete (https://agupubs.onlinelibrary.wiley.com/doi/10.1029/2005GC001148; https://academic.oup.com/gji/article/208/2/1173/2835566). Nevertheless, it is important to truly understand variations in the Earth's dynamo to integrate the geologic timescales of the carbon cycle with the longer-term evolution of Earth's climatic and ecosystem evolution, and we clearly are not there yet.

Second, let's talk Everest. As the tallest mountain on the planet at present, Everest gets a lot of attention. Climbers want to scale it (and sometimes tragically die in the process), and people use it as a universal comparator. The crater that marks the dinosaur-destroying Chicxulub impactor is slightly deeper than Mt. Everest is high. The tallest known mountain in the Solar System, Olympus Mons on Mars, rises three times the height of Mt. Everest. Aside from the fact that these comparisons to a mountain that few have actually visited are a bit pointless, they do speak to how unusual Everest is. But really, they speak to how uniquely high the entire Himalayan range is. It is so high and extensive, in fact, that it is easily photographed from the International Space Station Fig. 7.4

and generates its own weather system. This system, the Indian-Asian Monsoon, is responsible for collecting vast amounts of rain and snow

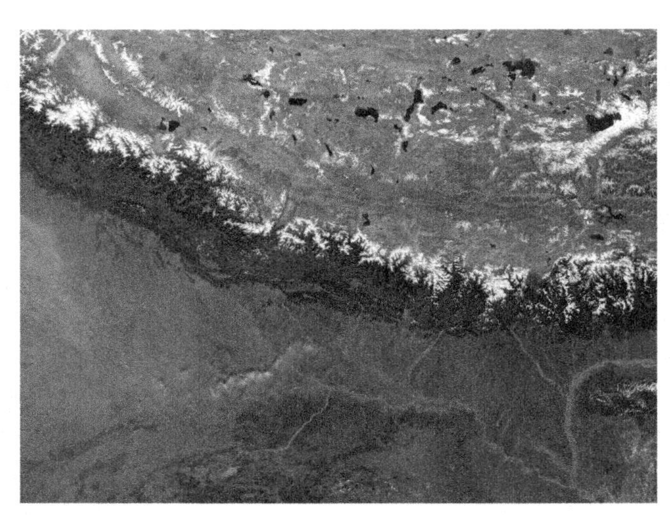

Figure 7.4 ***The Himalayas Mountain range as filmed from the International Space Station*** The Himalayas mountains, and the Tibetan Plateau behind them, are massive, weather-changing features on the Earth's landscape.

seasonally evaporated from the Indian Ocean, filling the mighty Indus and Brahmaputra Rivers, and producing the snow packs whose seasonal melting is critical as water sources for Himalayan mountain communities. Also, by effectively capturing all of the condensations that normally might have traveed across Asia and returning it to the Indian Ocean, it leaves downwind areas in a profound rain shadow and is responsible for the formation of the extensive Gobi Desert in China and Mongolia (https://www.sciencedirect.com/science/article/abs/pii/S0012825218307153). But like all geologic features on the planet, the Himalayas haven't been there forever, and actually represent a smaller example of previous record-holding mountain ranges, like the Appalachian Mountains of the eastern US, which at the peak of their geologic careers were as high as the Himalaya but are now eroded and collapsed to a small memory of their former glory. What really sets the formation of the Himalayas apart from some other mountain-building events such as that which drove the Appalachians is that the entire subcontinent of India was carved out of Antarctica and Africa over 100 million years ago and traveled as a continental mass north until it collided with Asia, beginning around 50 million years ago.

An India-sized chunk lying between Antarctica and Africa rent off during a rifting event about 125 million years ago, to become the so-called Indian Continent Fig. 7.5. (https://news.mit.edu/2015/india-drift-eurasia-0504)

This chunk of continental lithosphere eventually struck another giant continental mass—Asia. In a collision between oceanic and continental lithosphere, it is the ocean crust that ends up "losing," sinking down to be recycled in the deep Earth. If volcanic islands exist on the oceanic platform, they might get "scraped off" onto the edge of the continental lithosphere, but the ocean mass sinks nevertheless. But in a battle of colliding continents, neither is a clear favorite. When a continent-continent collision occurs, one of them tends to get stacked up onto another, making a double-thick layer of the continent that rises up into a massive mountain range. This is exactly what happened to form the Himalayas, with the first contact between the continental edges of India and Asia occurring about 50 million years ago and then proceeding to full-on contact and mountain formation beginning approximately 40 million years ago. Ultimately, the Himalayas continued to rise until they seem to have hit near their peak height sometime between 15 and 8 million years ago. This was the time when the Indian-Asian Monsoon kicked up into high gear. But as the mountains were rising, they were also being actively eroded. In fact, they were perfect erosion machines, given how much rain and snow they collected on their front range portions and their steep gradient—both factors cause high physical erosion rates.

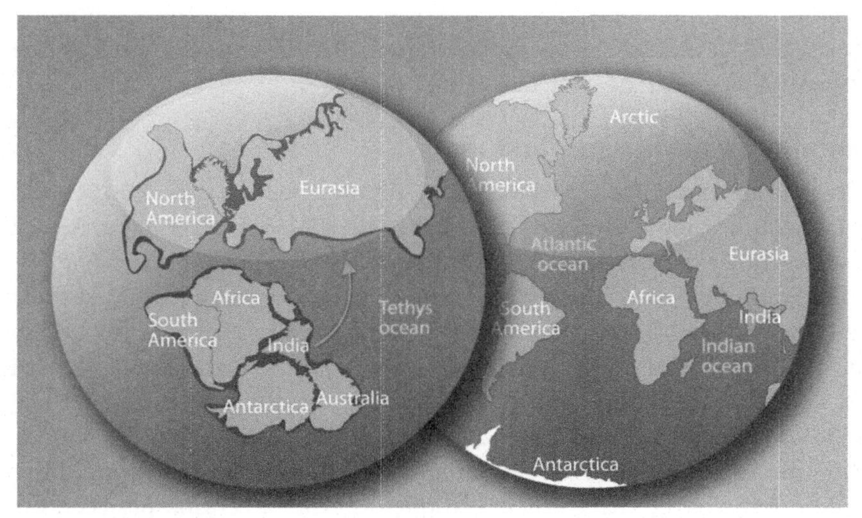

Figure 7.5 ***India Drifting into Collision with Asia*** The Indian plate rifted off of proto-Antarctica and collided with Asia starting about 40 million years ago.

It is this erosion that is a source of another long-standing concept that explains the ever-chilling Cenozoic climate—namely, that the process of weathering pulled so much carbon dioxide out of the atmosphere for so long that global temperatures dropped (https://news.climate.columbia.edu/2021/10/24/tackling-a-40-million-year-old-conundrum/). The isotopic evidence for this tectonic-climate concept is quite compelling. Not only can we clearly see the sheer volume of material eroded off the Himalayas and deposited in the Indian Ocean, but we can also see the impact that progressive erosion had on the marine strontium isotope values reflected in marine fossils. Continental lithosphere has very enriched strontium isotopic compositions, while oceanic lithosphere has very depleted strontium isotopic values. Since about 40 million years ago, marine fossils reveal a strong and very consistent trend toward more and more enriched strontium isotopic values recorded in their shell materials. Given that the only major mountain-building event over the Cenozoic has been the uplift of the Himalayas, and the Indian Ocean itself shows the eroded signature of this event as thick sedimentary deposits on the seafloor, the idea is that this strontium record shows the impact of extreme rates of Himalayan chemical weathering during this time. Although it is difficult to generalize long and complex processes like uplift, erosion, and climate change over a 40 million-year time frame, there is one particular aspect of this relationship that is well-constrained—the global impact that the ultimate uplift of the Himalayas had on terrestrial and marine life.

Mountains, monsoons, and mats

The "Himalayan uplift hypothesis" was proposed by Maureen (Mo) Raymo and colleagues in the early 1990s to explain the global cooling and increased glaciation that occurred through the second half of the Cenozoic. The hypothesis is largely based on the rapid rate of increase in the strontium isotopic ratio of marine carbonates since 40 Ma, cited as evidence for an increase in the rate of radiogenic (continental) strontium input to the ocean. Although the uplift and weathering of the Himalayan-Tibetan Plateau must have had profound effects on global ocean biogeochemical cycles, especially considering the large drainage area and unique geochemical characteristics of this region, some studies have called into question the direct link between the isotopic values and net chemical and sedimentological fluxes to the ocean (https://agupubs.onlinelibrary.wiley.com/doi/abs/10.1029/94PA01453?casa_token=a4ltyVfUm6EAAAAA:thZLv2UAtMtAK-gQwx9B8RwGqzSd

LE4Hzg1rFrLw7B3vMlizeJc9qdyfZVgR0ZbwWeSfiPa_iXi0vQ). A shortcoming of many of these studies is the attempt to link one or several of these oceanic geochemical indicators to records of climate change and tectonics throughout the Cenozoic or a large portion of it. As discussed throughout this book, the interplay between carbon cycling, planet reflectivity, and tectonics is always dynamic, and one simple explanation is not always adequate for explaining short or long-term changes in global temperatures. But with the uplift of the Himalayan-Tibetan Plateau, using a narrower time window and multiple distinct records of climate change and biological response can help to tie the story together.

This clearer record linking geologic, oceanographic, and climatic phenomena together revolves around the intensification of the Asian monsoon about 8 million years ago. This event links together continental weathering, geochemical and sedimentological fluxes to the ocean, biogeochemical cycles in the ocean, and the possible feedback on global climate. The Asian monsoon is a major climate system that currently affects global weather patterns. Massive warm rainfall events in late summer (from the Asian monsoon) combined with high elevations result in a disproportionate amount of chemical weathering occurring in the Himalayan-Tibetan Plateau Fig. 7.6. ; dissolved chemical fluxes in rivers draining this region account for about 25% of the global total, even though this region constitutes only slightly more than 4% of the global drainage area (https://www.nature.

Figure 7.6 ***Extensive River Flood Deposits in the Himalayas*** The Himalayas experience tremendous quantities of warm rainfall, causing high rates of erosion and chemical weathering. *Photograph by G. Filippelli.*

com/articles/359117a0). Oceanic and continental records suggest that the Asian monsoon may have intensified about 8 million years ago, possibly as a result of an uplift event in the Himalayan-Tibetan Plateau. This event had potential importance of this major climatic event on global cycles, including causing the "biogenic bloom" event documented in many ocean basins at this time.

Several records of weathering changes in the Himalayan-Tibetan Plateau centering on 8 million years ago have been extracted from the Indus and Bengal fans of the northern Indian Ocean. Trends in sediment flux rates to both the Indus and Bengal fans reveal a large peak that begins between 9 and 8 million years ago and persists to about 5 million years ago. Sediments associated with this peak are clearly Himalayan in origin, and the sediment input spike is probably related to the uplift and erosion of the Himalayan-Tibetan Plateau. Clay mineralogical changes and grain-size data from the Bengal fan show that chemical weathering intensity in the Ganges-Brahmaputra basin increased to a peak between 8 and 7 million years ago. The increased chemical weathering intensity in material weathering from the Himalayan-Tibetan Plateau is likely to have resulted in large increases in dissolved material. Thus, it is plausible that an uplift event produced more physical weathering products and intensified the Asian monsoon, causing these weathering products to undergo intense chemical weathering. IODP-supported research in the western Atlantic indicates a similar uplift and weathering history for the Andes, with uplift causing orographic precipitation from tradewinds and resulting in greatly increased sediment flux to the Atlantic at 8 million years ago (http://citeseerx.ist.psu.edu/viewdoc/download?doi=10.1.1.214.1993&rep=rep1&type=pdf). Increased rainfall in the Amazon basin may have combined with the higher amount of weatherable material in increased chemical weathering rates in this region. To explore the global effect of these regional phenomena, it is useful to examine several key oceanic geochemical records.

Geochemists often use the ratio of various isotopes or elements that are contained in marine fossil shells to reconstruct the chemistry of past ocean water and thus, by inference, various environmental conditions, including temperature, the rate of biological productivity, and even the rate of chemical weathering on land. The shells are handy because they often contain clues to when that organism lived, as well as in what marine environment (deep vs. shallow, free-floating, or fixed to the seafloor), depending on the type of organism the fossils represent. Several of these marine records reveal increased rates of elemental input to the ocean beginning about 8 million years ago. A global decrease in the oceanic geochemical ratios of the elements germanium and silica in marine diatom phytoplankton fossils over the past 35 million years is interpreted to reflect increased rates of total weathering, and particularly an increase in the rate of physical weathering versus chemical weathering. Superimposed on this overall decrease is a temporary reversal of this trend beginning at 8 million years ago and reaching a peak at about 6 million years ago. Fig. 7.7.

This short-term increase indicates a rapid increase in the proportion of chemical versus physical weathering. Several lines of evidence suggest that, if anything, river fluxes of silica from the Himalayan-Tibetan Plateau increased during this interval. First, a peak in sediment input argues for a greater net surface area of eroded material from the Himalayan-Tibetan Plateau, and thus higher chemical leaching rates of silica. Second, a global "biogenic bloom" event occurs during this interval, with microfossil accumulation rates peaking in many ocean basins during this interval, including both calcium carbonate-shelled organisms and silica-shelled organisms (such as diatoms). For all these accumulation rate records, a net increase in the global oceanic burial rate on these time scales is likely driven by an increase in the dissolved input of these elements from continental weathering. These records point to a relative intensification of chemical weathering at 8 million years ago, persisting for several million years. But the final piece of the puzzle is needed to wrap this up—where are the nutrients coming from to support and fuel all this marine plant growth?

On geologic time scales, the most important nutrient for fueling biological production in the ocean is phosphorus. Unlike the other major nutrient element nitrogen, phosphorus has no atmospheric phase and its delivery to the ocean is controlled strictly by the rate of weathering and transport of this element from the continents. Thus, the rate of delivery of this element to surface waters likely serves as the ultimate control on the oceanic biological pump, which in turn influences carbon removal from the atmosphere and burial on the seafloor. The late Cenozoic record of phosphorus accumulation rates in the equatorial Pacific reveals a large increase beginning at about 8 million years ago and peaking by 6 million years ago (Fig. 7.7). This increase must reflect a net increase in the rate of phosphorus input to the ocean, and cannot be accounted for simply by changing the distribution of phosphorus within the ocean. Furthermore, other records from the Pacific, the western Atlantic, and the Indian Oceans reveal this increase in phosphorus accumulation rates. This phosphorus peak coincides with high rates of physical and chemical weathering likely deriving from the

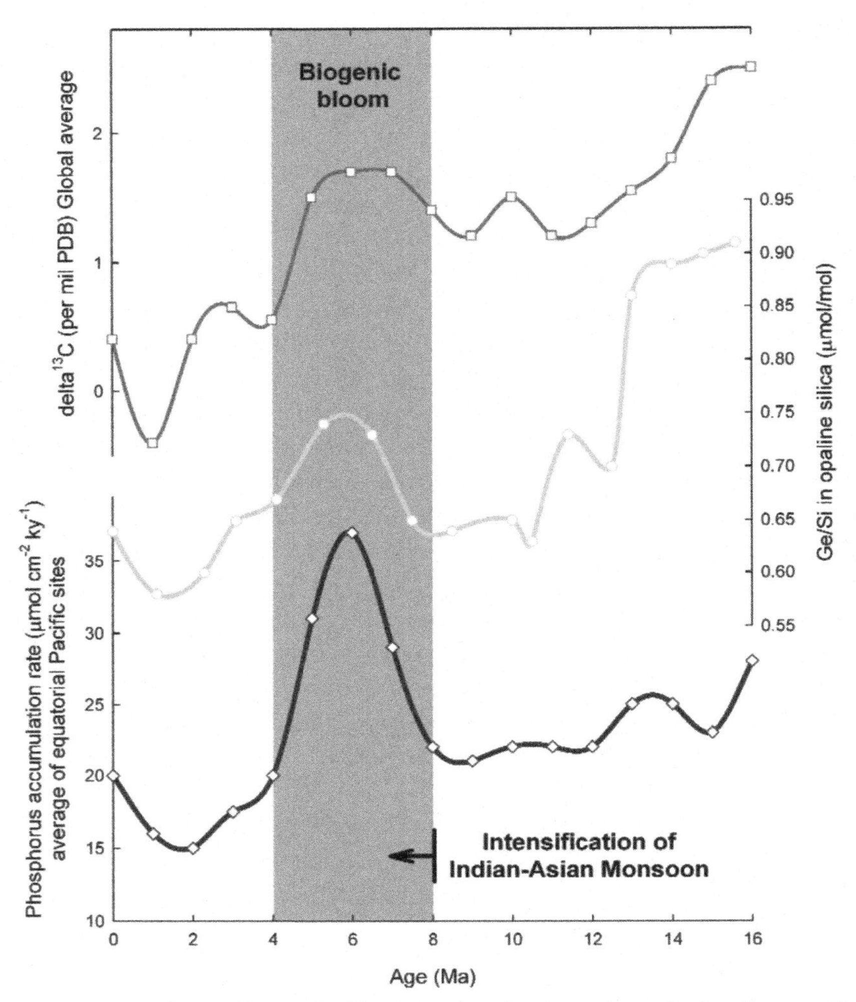

Figure 7.7 ***Geochemical Records of late Cenozoic Climate and Weathering Changes*** The peak from 8 to 4 million years ago in these records indicates a substantial increase in chemical weathering, nutrient supply, and biological production at this time, likely related to the onset of the Indian-Asian Monsoon system.

Himalayan-Tibetan Plateau (and possibly the Andes and Amazon Basin), as well as high biogenic productivity recorded in ocean sediments during this interval.

The mechanism for high phosphorus weathering rates during intense chemical weathering is revealed by a modern analog in the Amazon basin. In the lower Amazon, the transport-limited regime yields soil material that is nearly completely leached of phosphorus, a nutrient limitation that requires ecosystems in this region to extensively recycle this element Thus, an episode of high physical weathering rates in the Himalayan-Tibetan Plateau and other uplifting regions, coupled with high water fluxes and extensive chemical weathering in associated lowlands, would likely result in high phosphorus weathering rates and higher inputs of this nutrient to the ocean as reflected in the phosphorus accumulation rate trend.

The nutrient pulse to the ocean would be expected to increase net primary productivity and the export of organic carbon to the seafloor on these time scales. This hypothesis is supported by two observations. First, a broad expansion of the oxygen minima zone during this interval was caused by an increase in carbon export and oxidation at the sea floor, possibly driven by an increase in the net flux of nutrients to the ocean. Second, a transient positive shift in the ocean-wide carbon isotopic compositions coincides with the nutrient pulse (Fig. 7.7), indicating that the phosphorus nutrient input fueled more organic productivity in the ocean.

From an ocean ecosystem's standpoint, one marked difference toward the end of this nutrient pulse is the proliferation of diatoms and the formation of massive diatom mats. Diatoms are marine phytoplankton that makes their shells out of opal, which is a glassy crystalline form of silica. Diatoms require lots of nutrients, including phosphorus, which is being delivered in large quantities at this time, and silica, also with a plentiful supply given the enhanced weathering of the silicate rocks in the Himalayas around this time. Diatoms can grow extremely quickly, with doubling times of about 1 day, and tend to form dense mat-like floating structures in times of plenty. When the diatoms die, these mats sink down to the seafloor as sheets of nearly pure opal. Thus, diatom mats in ocean sediments are a good indication of high organic productivity in the overlying water. An interesting aside is that these mat layers are notoriously hard for scientific ocean vessels to recover from seafloor sediments—they often appear as tough layers interspersed with softer layers of clays or calcium carbonate-rich sediments from the other major marine phytoplankton—coccolithophoridae. When coring tubes are used to collect intact marine sediments from the seafloor, the diatom mats resist the sharp core edges until enough pressure is delivered to break through them, which invariably is too much pressure to capture intact records of the softer interspersed sediments. One of the many challenges of science!

A significant increase in the rate of chemical weathering during this time may have had an impact on the long-term carbon cycle. First, the direct effect of the weathering of silicate minerals (combined with calcium carbonate sedimentation in the ocean) is an important long-term sink for atmospheric carbon. Second, an increase in organic matter production and burial in the ocean due to an increase in the nutrient flux is an additional sink for atmospheric carbon. Most models of Cenozoic climate invoke just such long-term carbon sinks for the observed deterioration of global climate. But the continental weathering event from 8 to 4 million years ago appears to have had a little immediate effect on climate, even if a long-term drawdown of atmospheric carbon dioxide associated with this weathering event did tip the climate balance from its previous stable state and into colder conditions toward the end of the Cenozoic. The crystal ball of science is murkier than usual during this interval, and will only be cleared up by more work on productivity and weathering records.

Ocean gateways open, and close

The current configuration of continental lithospheric plates on our planet is one near the maximum of potential geographic separation. Where a short 150 million years ago you could take one step and have gone from Boston to Casablanca, the distance between the two is now 5517 km (3428 miles), which would be more like 7.7 million steps! A good look at the globe shows that nearly all of our continental plates are currently separated from each. This is in sharp contrast to 250 million years ago, around the beginning of the Age of the Dinosaurs, when all the continents were aggregated into one big landmass, called Pangaea. To be clear, even then most of the plates themselves were separated from each other by invisible "seams" that were easily torn asunder by tectonic processes, and in fact, our history since then shows them doing just that. The only significant areas of new continental aggregation are the aforementioned Indian subcontinent, now attached to the Eurasian Plate, and much of the far western US, which has a repeated history of aggregation of volcanic arcs and other microplates that were "scraped off" of the oceanic plate that is largely subducting under the western US. But since the beginning of the Cenozoic, a large number of major and minor tectonic movements have significantly shaped ocean circulation, weather patterns, and the global temperature gradient.

To expand on the brief summary of Cenozoic tectonic shifts presented in the earlier part of this chapter, it is helpful first to understand modern ocean circulation patterns. Why start with the motion of the ocean? First, a significant amount of heat and moisture transfer on the planet is due to ocean circulation patterns. The ocean doesn't move as one monolithic mass, but rather there are distinct vertical zones in the ocean that move based on various forces, and are relatively separate from each other (as much as one water mass can be considered separately). First, the surface layer, confined to about the upper 200 m of the water column, moves from the action of wind. The wind literally moves the surface ocean around and creates the surface currents that are responsible for 61,280 Nike shoes, which fell off a storm-wracked ship steaming from Seoul to Seattle, winding up on the beaches of Oregon and Washington a year later. This is actually a fascinating story, perhaps one of the rare positive aspects of marine pollution as it led to a very sophisticated natural experiment where citizen scientists helped to trace the rate and direction of surface ocean circulation patterns (https://magazine.washington.edu/feature/how-61000-floating-nikes-helped-an-oceanographer-find-his-calling/). Next in depth are the Intermediate Ocean currents, which display variable thicknesses, move more slowly than the surface ocean layer, and are a sort of mixture of surface and deeper ocean systems.

Finally, deep ocean currents, whose patterns and flow directions are dictated by the configuration of continents and the density of water that comprise them, making up the vast majority of ocean waters by volume. Deep ocean circulation is quite slow—if you could somehow manage to get Nikes to sink and move with deep ocean waters, a shoe spilled in the youngest deepwater in the North Atlantic Ocean would not emerge on the beaches of California until about the year 4200 AD. Deepwater currently forms in the North Atlantic Ocean because it is cold and extremely dry, making the surface water there denser than the surrounding water and causing it to sink Fig. 7.8.

With nowhere else to go, it makes its slow way from Iceland down to near Antarctica where it contributes to and is replaced by Antarctic Bottom Water, which then sweeps around the Southern Ocean surrounding Antarctica and circulates into the deep Indian Ocean and eventually into the Pacific Ocean. The sinking surface water in the North Atlantic is replaced by Intermediate Water from the very salty Mediterranean Sea and by water that originated in the Gulf of Mexico region but is swept north due to Coriolis forces that dictate surface circulation on a revolving sphere. This global system of deepwater formation and surface water replacement was termed the Ocean Conveyor Belt (Fig.; https://www.nasa.gov/topics/earth/features/atlantic20100325.html) by the grandfather of many

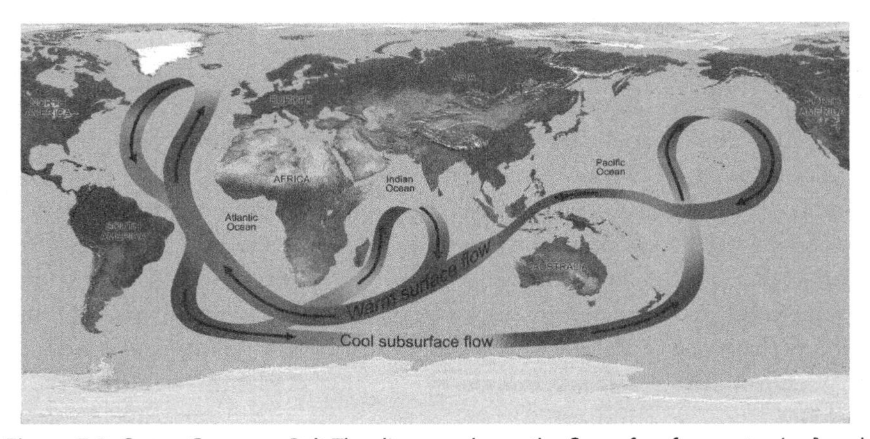

Figure 7.8 ***Ocean Conveyor Belt*** The diagram shows the flow of surface water (*red*) and deep water (*blue*) around the globe. *From NASA.*

fields of paleoclimate, the late Wallace (Wally) Broecker from the Lamont-Doherty Earth Observatory of Columbia University.

Wally recognized that this system of global ocean circulation changed significantly through geologic time, both as a function of where on the planet denser surface water could be formed and by the configuration of circulation pathways within ocean basins. For example, the Gulf Stream currently moves a great amount of water—the equivalent of 150 Amazon Rivers—and exports a significant amount of tropical water that warms Northern Europe. Would you rather be in Stockholm or northern Saskatchewan in winter? Well, they receive the same amount of sunlight, but the average January temperature of Stockholm is almost 30° warmer than northern Saskatchewan owing to the moderating effects of the ocean. Just 20,000 years ago, however, the great glacial polar ice sheets advanced all the way over the area of the North Atlantic where deepwater is currently formed, causing a significant southward deflection of the Gulf Stream and also causing the whole global ocean deep circulation system to back up. At this time, the choice of winter in Stockholm or Saskatchewan would be a toss-up.

Along with its role in transporting heat around the planet, deepwater circulation is also largely responsible for providing new nutrients to fuel surface ocean biological productivity. Where deepwater upwells, largely at the end of its circulation path on the margins of continents and in equatorial and polar areas where surface circulation patterns diverge and make way for deepwater, this water is typically relatively old. It has also accumulated nutrients from the degradation of organic matter that rains down to the depths as the deepwater mass slowly makes its way through ocean basins, where it is too dark to support the phytoplankton that would otherwise consume it. These nutrients feed the vast blooms of algae, and the zooplankton and fish that consume them, that are found in these upwelling regions, making them the most productive fishery regions on the planet. Ironically, these regions have the lowest biodiversity as the oversupply of nutrients results in a very leaky and inefficient food web. This contrasts with the low upwelling regions that can be found in vast areas of the ocean and which promote highly diverse, efficient, but low productivity ecosystems such as coral reefs.

Consider now the circulation pattern that was extant at the beginning of the Cenozoic. At that time, the Indiana subcontinent was still making its way up through the Indian Ocean to Eurasia, and there was an ocean gateway north of it. Africa had not yet collided with Europe, and the restricted Mediterranean Sea didn't exist. The Isthmus of Panama was not a closed continental bridge, requiring malaria-riddled excavators to carve a shipping canal through, but was instead a wide-open seaway. The equatorial ocean region was largely free of circulation obstacles, and this globe-circulating surface ocean system was termed the Tethys Ocean or Tethys Seaway. It was quite shallow at various times in its history, and before the beginning of the Cenozoic it was likely large enough, hot enough, and shallow enough to become a "factory" of super dense salty water, which likely was a dominant deepwater source for the ocean at this time.

Also at the beginning of the Cenozoic, Antarctica had not yet separated from South America, and thus the modern polar-encircling circulation pattern that constitutes the Southern Ocean did not exist. The modern Southern Ocean is a significant component of deepwater formation and circulation, owing to the fact that it circulates continually around Antarctica in a polar circle that is quite cold, and the only water that mixes with it is deepwater that arrives from the already-cold North Atlantic source. The Southern Ocean is basically an isolated thermal wall, and in this way is responsible for the extremely cold temperatures that exist in Antarctica today. At the beginning of the Cenozoic the Southern Ocean didn't exist, and the modern situation in the surface North Atlantic Ocean, where warmer tropical waters are exported through surface ocean circulation, was the norm around Antarctica. It is thus no surprise that even at its extreme southerly latitude, Antarctica at that time was relatively warm and contained some ecosystems and organisms, such as crocodiles, that would be considered downright tropical today.

Through the Cenozoic, these ocean pathways of heat and nutrient transport slowly shifted. Antarctica began rifting away from South America about 40–30 million years ago, and the seaway, which opened between these continental masses, the Drake Passage, became deep enough to begin the ocean circulation pattern, called the Antarctic Circumpolar Current, that comprises the Southern Ocean and thermally isolated Antarctica, which in turn began forming its first ice sheets on its way to being the white ice-covered continent that it is today. Meanwhile, on the other pole, the Arctic Ocean basin had a very different history. In the earlier portion of the Cenozoic, the Arctic appears to have been relatively "fresh" and warm, based on evidence of fern fossils, and it certainly received significant riverine input from the Canadian and Siberian arctic region (https://tos.org/oceanography/assets/docs/19-4_moran.pdf). By the middle of the Cenozoic, however, the Arctic Ocean started showing evidence of glacial and iceberg input, and strong evidence for significant floating sea ice started shortly after that. Effectively, the Antarctic ice sheet and extensive Arctic sea ice began at a similar time, and in the case of the Arctic, the increased albedo from ice reflection exerted a profound regional cooling effect. Although our understanding of the full climatic history of the Arctic is very incomplete (it is very difficult to get scientific seafloor coring expeditions in this area due to the sea ice cover), it seems that the Arctic waters themselves became moderately isolated from the rest of the world's oceans after the middle part of the Cenozoic, with the North Atlantic gateway and the Bering Sea being the main, albeit limited, circulation arteries.

During this time the Tethys Ocean sequentially closed, creating dramatic features in the process, such as the Alps, where abundant and beautiful marine fossils can be found, long before deposited on the Tethys seafloor and now exposed on the mountain slopes. The Tethys closure started the process of ocean isolation. First, the seaway between the Atlantic and the Indian Ocean closed, replaced by the restricted hyper-saline Mediterranean Sea and causing the uplift of much of the Middle East. As an aside, portions of the Tethys Ocean were also quite biologically productive, resulting in the significant seafloor accumulations of organic matter, which eventually became the oil fields and phosphate deposits that have enriched this region. The equatorial closures continued, with the final gateway between the Atlantic and Pacific Ocean basins closing with Panama uplift about 4 million years ago. This latter closure led both to the increasing divergence of marine flora and fauna on either side of Panama and also the increasing terrestrial migration of organisms across this land bridge.

Horses, and grasses, and silica, oh my!

Teeth provide an interesting insight into the evolution of organisms. Take shark teeth, for example. Sharks are one of the longest-lasting animal groups in existence, having emerged as a distinct group around 400 million years ago. So obviously, sharks were pretty well-designed from the get-go, including, it seems, their orthodonture. Sharks evolved to continuously make new teeth, which are lined up behind the older ones like soldiers in formation, because their style of hunting depends on both a sawing action when the biting action occurs and also inward-facing teeth angles to ensure that the prey doesn't getaway. The stresses and strains of this eating style results in lots of broken teeth during a shark's lifetime. And you have never seen an old toothless shark because of the new sets of teeth always standing in wait for their turn to chomp.

Humans are quite different. "That fellow is getting long in the tooth" or other variations on this phrase, actually have nothing at all to do with judging how old a person is, because we only grow two sets of teeth, and their size is fixed. As human beings have lived longer, we are feeling the oral pain of this evolutionary approach to teeth. We have a set of initial teeth that are pushed out by our second set of "permanent" teeth. Really, though, that second set is designed to replace primary teeth missing due to decay or breakage, and thanks to modern dentistry and nutrition rarely is this the case (barring the occasional "string tied to a doorknob" move). And when it comes to wisdom teeth, we are completely underdesigned to fit modern diets. These typically come in between 18 and 20 years of age, and they are supposed to be coming in to fill gaps left from decay or loss of so-called permanent teeth. In this way, evolution + nutrients/dental care = happy orthodontists! On the flip side, we are also woefully underdesigned to fit modern human lifespans, and thus by the age of 60, people have lost an average of a third of their teeth, and more than half of people who are 85 and older have no teeth left at all! A little shark-type evolution would be welcomed at the senior center.

Horse teeth, however, are a completely different story, and tell a history of a geochemical arms race between plants and beasts. For horses, "long in the tooth" literally refers to an old horse, because horses' teeth continue to grow throughout their entire lives. Horse tooth evolution is a fascinating topic, but first, what about just plain horse evolution? Horses have been around since about the PETM interval 55 million years ago, although their evolutionary appearance likely had nothing to do with this particular climatic event. It would have been a challenge for nonpaleontologists to recognize this first group, as they were closer to the size of foxes and dined mainly on fruits, as evidenced by short flatter teeth. Horses slowly grew larger and their diet changed in response to increasingly cooler climates that began to prevail around 33 million years ago, which brought about a shift in ecosystems toward leafy plants. Then, a shift in horse diets occurred, starting around 18 million years ago as evidenced by tooth structure, to teeth that could withstand greater abrasion. This trend continued until about 5—4 million years ago, when the modern horse evolved from these ancestors, with a clear evolutionary advantage—continually growing teeth! But why the need for this outlay of energy, and how did it relate to ecosystems at the time? It all comes down the silica, grasses, and a new way of doing photosynthesis.

Sometime in the middle part of the Cenozoic, plant leaves started to become coarser, and less easily chewable and digestible. This was an evolutionary response to grazers like horses, much like thorns are protection for rose bushes. The adaptation here is that grazer teeth would wear down from the toughness of the leaves, and make them less able to, well, graze. They did this with a clever trick—by absorbing soil-based silica and incorporating some of it into their tissues, they are effectively building mineral abrasives into their leaves, which wear down teeth. These silica structures are called "phytoliths," which roughly translates as plant rocks. Phytoliths came on the evolutionary scene in the middle part of the Cenozoic at the same time that the climate was cooling. These plant mineral components are well preserved in lake and ocean sediments, and a special breed of paleontologists who examine microscopic organisms (micropaleontologists) are able to reconstruct at least some aspects of terrestrial ecosystems from the preserved record of phytoliths. Different plants make different morphologies of phytoliths, thus leaving behind a record of not only how many plants there were on the landscape, but which plants. It has been proposed that the plant evolution of phytolith building might have fundamentally altered the terrestrial silica cycle because more silica becomes dissolved during this process and can make its way to lakes and oceans. This is perhaps one reason why phytoplankton that makes their shells out of silica, such as diatoms, began dominating lakes and oceans toward the latter part of the Cenozoic.

Horses started forming teeth that had a capacity for continued growth to counteract the evolution of phytoliths Fig. 7.9.

This seems to have started a horse-plant arms race, which culminated in the proliferation of grasses around the world. This proliferation occurred in that same 8–4 million year ago timeframe of peak Himalayan uplift and weathering, and may indeed have been caused by it. Grasses are not only super-phytolith producers (https://besjournals.onlinelibrary.wiley.com/doi/10.1111/j.1365-2656.2006.01082.x#:~:text=Silica%20in%20the%20leaves%20of,1985), but they also exploit an entirely different way to convert sunlight into sugar. If you have ever rolled around on a grass lawn or walked through a field of tall grass, you have experienced super-phytoliths firsthand. Modern grasses are 2%–5% silica (dry mass), and that silica is present in the form of phytoliths, and these sharp minerals cause the skin microabrasions that are so uncomfortable after grass encounters. Grasses also utilize the C4 photosynthetic pathway, which is more complicated than the standard C3 pathway and thus more metabolically expensive but is highly efficient in drier conditions with lower concentrations of atmospheric carbon dioxide. It is no surprise then that the C4 plants proliferated toward the end of the Cenozoic, as atmospheric concentrations of carbon declined, temperatures dropped, and the atmosphere became drier (a consequence of lower temperatures). Although common turf grass that you need to mow every week in the summer is a C4 plant, so too are many of the grain plants, like corn and sorghum, that we utilize today to support our human populations.

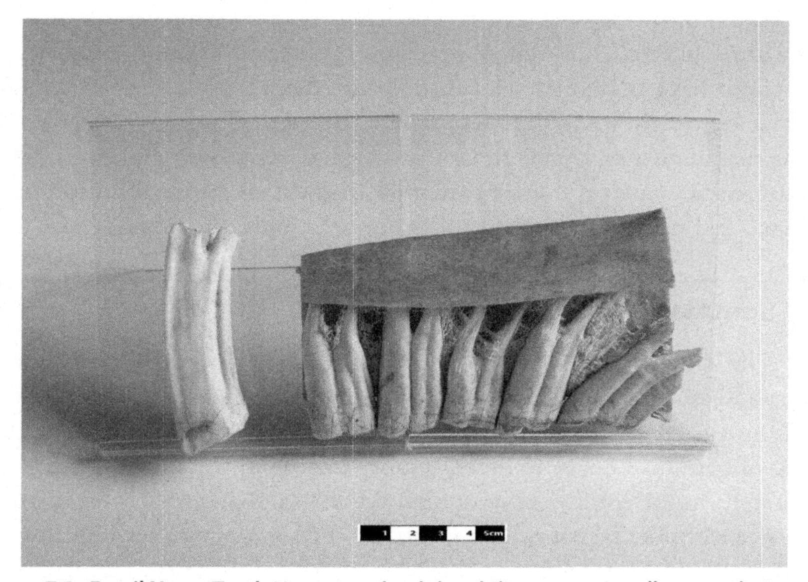

Figure 7.9 ***Fossil Horse Teeth*** Horses evolved the ability to continually grow their teeth to counteract the erosional effects of phytoliths from grasses in their diet.

Horses ultimately evolved to eat grasses almost exclusively and could do so given that they were armed with teeth that never wore down. There was little competition for grasses and given their proliferation, they were an abundant globally-distributed food source. And horses are not the only organisms that figured out the continuously growing tooth thing—rabbits and rats, rats and rhinos, cows and camels, they all can get "long in the tooth," particularly if you replace some of their traditional abrasive diets with nonabrasive foods (like oats in the case of horses). One of the critical design features for horses, and all hoofed creatures (so-called ungulates), is that eating grass is not enough—they need to be able to digest it. Here is where their unique digestion comes in. With multiple stomach chambers and the typical behavior of chewing, swallowing, and regurgitating consumed greases to digest again, horses and other ungulates have evolved to slowly but surely to break down the cellulose-heavy composition of grasses and other tough plant parts. Cellulose digestion is a challenge, one that humans and many other animals can't overcome because our digestive systems are filled with air, and all of the oxygen that comes with it. The unique microbes that can digest cellulose require an oxygen-free environment, and the multiple stomach chambers of ungulates allow one of those chambers to be oxygen-free and full of cellulose-chomping microbes. One consequence of this is that the microbial byproduct of consuming cellulose is the gas methane, and as discussed earlier, methane is a powerful greenhouse gas in the atmosphere. Thus, when ungulates evolved earlier in the Cenozoic, they might have shifted the global methane balance a bit. This shift might not have been significant, but modern livestock agriculture has tilted the system so that ungulates, largely those that we raise in captivity and consume in large quantities (cows and sheep) are now a significant source of atmospheric methane and play an unwanted role in driving global warming.

Summary

The Cenozoic is "our" time, the Era of mammals (https://www.panmacmillan.com.au/9781529034226/), and climate change throughout the Cenozoic has dominated mammalian evolution. Some call the Cenozoic a time of continual "climate deterioration" owing to the shift from Greenhouse to Icehouse conditions, but this holds the value judgment that warmer is better and colder is worse. The Cenozoic did see a significant shift toward climatic zonality, particularly in the latter portion of the Era. This is due to relatively unique circumstances of having an Antarctic circulation pattern restricting tropical heat transport and the Arctic Ocean that opened up, and in doing so provided the capacity for seasonal and multi-annual sea ice accumulation and thus higher polar albedo. Both polar regions had perfect ice-growing conditions, although for very different reasons. And as the atmospheric carbon dioxide concentrations kept declining during the Cenozoic, causing a decreased greenhouse warming, grow ice they did, causing global temperatures to go down even more due to enhanced albedos.

Of course, the planet doesn't care about terms like "climate deterioration," it simply adjusts its thermostat based on the amount of greenhouse gasses in the atmosphere and the balance of incoming to outgoing solar radiation, a result of global reflectivity. And in a sense, the "climate deterioration" is relatively unimportant to the mammals that have reigned during this time. Warm-blooded organisms certainly do expend more energy when conditions are colder, but otherwise, they can function much better than their nonthermal regulating counterparts. Thus, even with the enhanced zonality, mammals are found in low and high latitudes alike and have developed myriad adaptations to fit their climatic homes. In equatorial regions where temperatures regularly exceed the internal body temperatures of mammals, they developed sweating or panting approaches to cooling. In higher latitude regions where temperatures can be significantly below internal body temperatures, they evolved thicker coats of fur and hair to reduce heat loss.

Even among cousin organisms, the morphological distinctions between warm climate and cold climate dwellers are obvious. Take for example the general family Elephantidae. The African elephant has nearly no fur and large, high surface-area ears that act as heat radiators to enhance cooling in the hot climate. The now extinct Wooly Mammoth, in contrast, roamed the tundra landscape of the last Ice Age and had extremely dense fur, a thick subcutaneous layer of fat, and small ears to reduce heat loss. This type of climatic adaptation is just one of the examples of the general environmental flexibility of mammals. To take Elephantidae analogies even further, an earlier cousin of the African elephant diverged and took its grazing behavior to the water, eventually becoming the modern-day manatees and dugongs that are seen throughout tropical waterways.

Humans took these climate adaptivity tools even one step further. With intelligence that allowed for tool making and technologies, they harnessed fire and donned furs (and parkas) for warming. These adaptations meant that they were relatively unaffected by ambient temperatures, especially when they started using shelters as the thermo-regulating system. Once indoor heating and cooling came about, we could comfortably wear t-shirts in winter in Minsk and sweaters in summer in Miami, so long as we were inside. Ironically, to do this we resorted to the main ingredient in the earth system that has always influenced global climate-carbon—and in doing so, have changed the trajectory of planetary climate in such a way that we are at risk of Minsk feeling a lot like Miami.

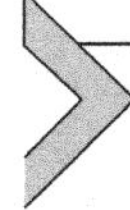

Did you know that?

Scientists have posed the question of whether we are the first "civilized" creatures who have inhabited the Earth

This issue is an important one to at least theoretically consider, as the duration of an advanced civilization's persistence on a planet is one critical factor to consider in our search for life on other planets. The Drake Equation is one construct frequently used to estimate how many advanced civilizations there might have been in the universe since the Big Bang over 13 billion years ago. There might be millions of advanced civilizations that have bubbled up, reached their peaks, and then faded, through the attrition of time and conflict, like the Chinese Dynasties and the Roman Empire, or through some global conflagration or massive meteor impact. Our industrialized civilization, marked by the beginning of the production of energy through the burning of fossil fuels, has only been around for less than 200 years, and yet we have left a huge mark on the planet. If we persist for another 200 years, or 2,000, it pales in comparison to the extraordinarily long period of time that the Earth has been a planet. So even if those millions of advanced civilizations in the universe persisted for 10,000 years (perhaps an overfly generous allowance), that still means that there might be only one or two advanced civilizations active at any given time across the ENTIRE universe.

A search for a current advanced civilization somewhere else in the cosmos is limited by the vastness of space, but would we even be able to identify one on our own planet in the depths of geologic time? This was the question posed by scientists Gavin Schmidt and Adam Frank—what is the signature in the geological record that is being produced by our current advanced civilization, and would we be able to detect it in the past (https://www.theatlantic.com/science/archive/2018/04/are-we-earths-only-civilization/557180/https://arxiv.org/abs/1804.03748)? For example, was the Paleocene-Eocene Thermal Maximum (PETM) 55 million years ago the result of some earlier civilization combusting a massive amount of fossil fuels like we are, leaving behind a brief period of high atmospheric carbon dioxide, warming, ocean acidification, and extinction? Our romance with burning fossil fuels is likely fleeting, but we are also leaving a series of non-climatic signatures on the planet. High amounts of nutrient enrichments from fertilization, the mining and use of unique metals, the application of novel and nonnatural chemicals like pesticides, herbicides, PFAS, and CFCs, and a thick layer of plastics that may preserve quite well in the geological record.

Would we be able to see any of those if they were similarly part of the PETM civilization? It is actually debatable, as the layer of civilization would be so small, especially if short-lived, compared to the typically slow processes of geological accumulation. But perhaps a better question is why haven't we looked for them? As Schmidt and Frank themselves admit, this is largely a thought experiment, and we do not have evidence of previous advanced civilizations on our Earth, whether 55 million years ago or 555 million years ago. Nevertheless, the fleeting nature of civilizations might be like lonely lightning bulbs at the end of the mating season, with a flash here and there in the cosmos but nobody there to answer the call.

Marsupials were saved by geology?

Marsupials are only endemic to Australia, the South Americas, and part of Indonesia. They are part of the class of mammals but diverged from the "placentals" about 160 million years ago. As the super-continent of Gondwana split up and the continent became more dispersed, marsupials themselves became isolated to a few continents, and their isolation may have been their salvation. Importantly, the marsupials found in Indonesia are mostly recent immigrants from Australia, and almost all marsupials in the Americas are in South America, with the exception of opposums. This geographic isolation might not be a coincidence, but instead, one forged by plate tectonics.

The Australian Plate is connected across the Straits of Carpentaria with parts of Indonesia. During glacial intervals, when sea levels are low, there is a land bridge between the two landmasses, but otherwise is one of the most isolated of the continents, with the exception of Antarctica. Meanwhile, South America is only connected to the rest of the Americas by the Isthmus of Panama. As discussed earlier, this connection is a relatively recent one by geologic reckoning, occurring only about 2—4 million years ago. The extended intervals of isolation for these two continents meant that marsupials were able to evolve largely in isolation from placental predators.

This has resulted in marsupials filling in most of the placental mammal niches in Australia and South America thus diversifying. It is not until humans started translocating species all around the globe that the marsupial world has been encroached upon by the placental mammal, and the declining number of marsupials (https://onlinelibrary.wiley.com/doi/10.1111/geb.12088) speaks to who has come out the loser in this development.

CHAPTER EIGHT

Climate and humans

Introduction

We humans, as the species homo sapiens, have not inhabited this planet for very long. Just 0.04% of Earth's history includes us walking the planet. For such a short tenure, we have had a huge impact, including reshaping Earth's surface and its ecosystems, changing climate more rapidly than it has ever changed before, acidifying and warming the oceans faster even than the shockingly rapid shift that occurred 55 million years ago. Intriguingly, we homo sapiens evolved to survive, and thrive, through the up-to-then dramatic climate shifts that have occurred through glacial and interglacial variations. Much as the environment shapes ecosystems and ultimately influences evolution, environmental changes triggered by climate substantially bent the arc of human evolution.

The earliest hominids like Lucy Fig. 8.1, an ape-like Australopithecus afarensis found in East Africa, couldn't survive climate shifts that changed the landscape from forest to savannah and then back to the forest, and went extinct about 3 million years ago after surviving for 900,000 years. Our most recent planetary coinhabitant, the hulking Neanderthal, evolved to have large sinus cavities, which allowed them to breathe comfortably in the chilly air of Ice Age Europe but eventually met its demise from the extremely high caloric needs to fuel its muscular bulk. Many northern Europeans have Neanderthal DNA tracing back to cross-breeding with homo sapiens, but nevertheless, the modern version of hominid is us. We are shaping the global climate, and due to technologies are not as vulnerable to the very climatic shifts that we are creating as our ancestors were.

This chapter covers two very broad phases of climate and human life: (1) how has climate shaped landscapes and human evolution, and what role did early pre-Industrial humans play in landscape and climate changes?, and (2) what are we doing now, and likely will continue to do in the future, to shape the fabric of climate and life on our planet? The first topic relates very strongly to humans as "hapless victims of climate change," while the second casts humans as "willful and culpable engineers of planetary changes" far beyond the scale and rate that Earth has seen for the past 4.5 billion years.

Climate Change and Life
ISBN: 978-0-12-822568-4
https://doi.org/10.1016/B978-0-12-822568-4.00009-2

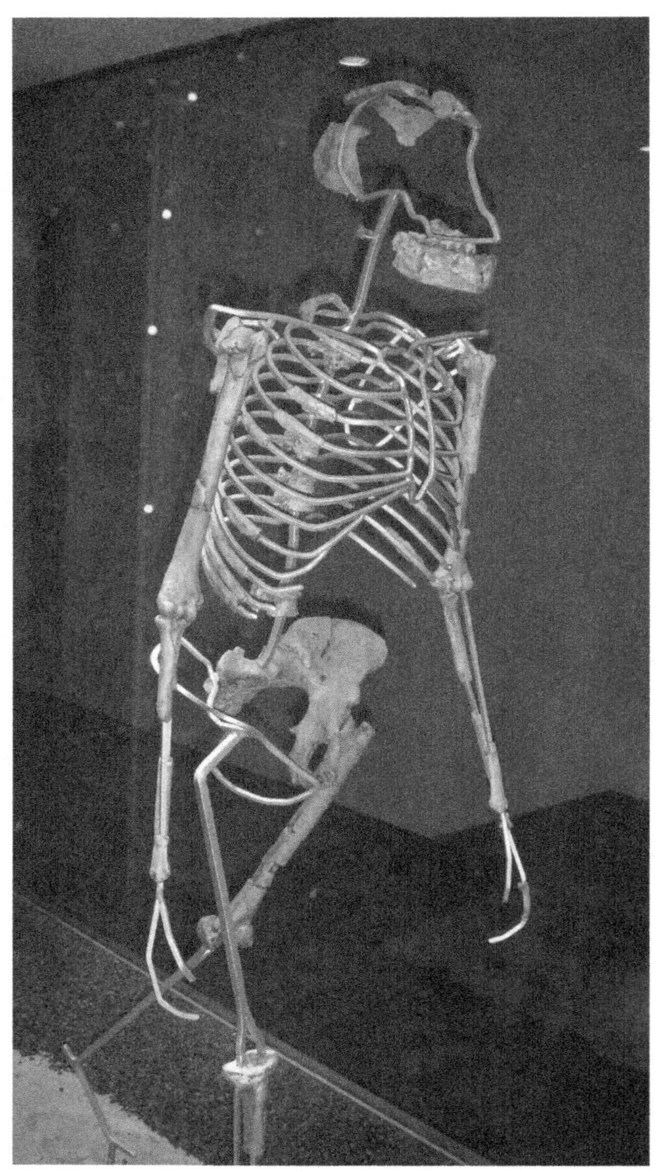

Figure 8.1 ***Skeleton of Lucy*** A remarkably complete fossil of an ape-like female Australopithecus afarensis found in East Africa was dubbed "Lucy" after the Beatles song "Lucy in the Sky With Diamonds," which apparently was playing on near loop in the expedition camp during the excavation.

This is not intended to be a disaster list of impending planetary doom, but rather a science-based understanding of what our planet will look like in the future, and how organisms and ecosystems will survive and evolve in the new Anthropocene—the Age of Humans.

Evolutionary processes and hominids

For almost 150 years, the general paradigm of evolution being driven by environment and environmental change has held strong. Charles Darwin published his "On the Origin of Species" in 1859 Fig. 8.2. Behind the theory of evolution that he offered in this book is that natural selection tends to favor the survival of individuals and species who are most adapted to the environmental conditions at the time and that the long course of time will yield family trees of ancestral species with each new branch forming as the result of an evolutionary step. The concept of natural selection shaped fields far outside of biology, including business and military practices where it underpinned "survival of the fittest" concepts. But scientific challenges to the theory came early, largely revolving around the fact that, in Darwin's time, there was no field of genetics, nor knowledge of DNA and genes as the biological roadmaps for reproduction and function. So, no driving factor was laid out to indicate that the breaks in the tree were caused by genetic mutations that conferred advantages to the individual and species that had them. I say "individuals" here because it is a mutation on an individual's gene that is advantageous to greater survival, and that individual then spreads that mutation to the entire species. Of course, there are likely millions of times where that individual's advantageous gene does not get carried on or does so only locally and without broader transmission to the global species. Nevertheless, it is this mutational pathway that has been essential to the long history of life on this planet.

Evolutionary biology did not stop with Darwin, and indeed one of the key refinements to the theory came from the Harvard paleontologist Stephen Jay Gould. Before Gould, there was largely an assumption that environments change slowly, so rates of evolution would similarly progress slowly and methodically. But Gould, trained as a geologist and aware of extreme climatic and environmental shifts reflected in the fossil record, argued that evolution occurred in fits and starts. This concept of "punctuated equilibrium" called for periods of climatic stability, where evolutionary rates (mutations) were generally low, interspersed with significant environmental shifts when most of the evolutionary mutations occurred. Whether

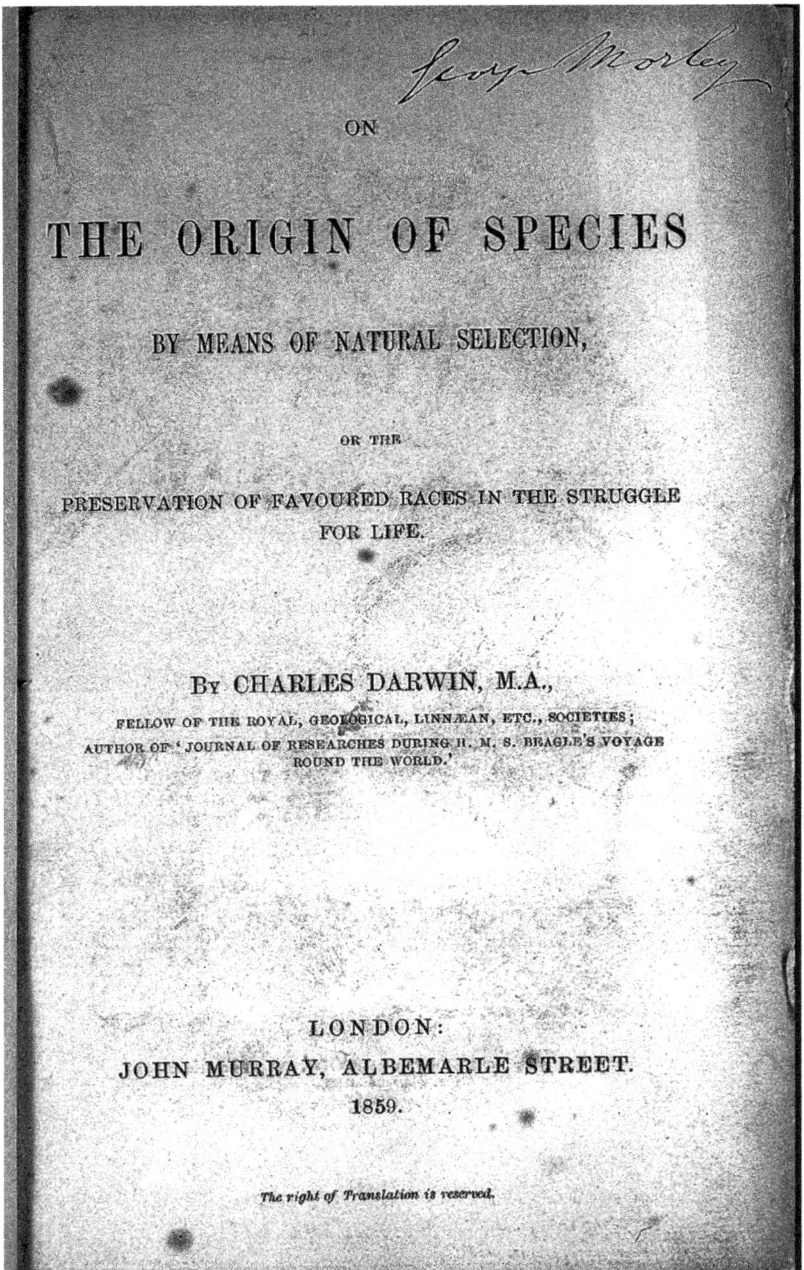

George Morley

ON

THE ORIGIN OF SPECIES

BY MEANS OF NATURAL SELECTION,

OR THE

PRESERVATION OF FAVOURED RACES IN THE STRUGGLE FOR LIFE.

BY CHARLES DARWIN, M.A.,

FELLOW OF THE ROYAL, GEOLOGICAL, LINNÆAN, ETC., SOCIETIES;
AUTHOR OF 'JOURNAL OF RESEARCHES DURING H. M. S. BEAGLE'S VOYAGE ROUND THE WORLD.'

LONDON:
JOHN MURRAY, ALBEMARLE STREET.
1859.

The right of Translation is reserved.

Figure 8.2 ***Original Cover of "On the Origin of Species" by Charles Darwin*** Published in 1859, this book slowly rocked the scientific, and theological, world, providing a roadmap for what later became the field of evolutionary biology.

these mutations nearly instantaneously spread in a population (Macromutation) or occured more gradually over a short time period of thousands of years is difficult to constrain because of the limitations in the geological record Fig. 8.3. but regardless, this concept of punctuated equilibrium is consistent with our understanding of genetic mutations and the spread of these mutations across a species.

Lest you assume that evolution always follows an "onward and upward" trajectory, consider photosynthesis. As outlined earlier in this book, photosynthesis evolved several billion years ago and involved an Archaea inadvertently absorbing an algae-like bacterium and miraculously surviving this infection. Even more miraculously, perhaps, passing this weird hybrid down to the next generation, and the one following. Perhaps this engulfing process occurred millions of times over millions of years, with many of the infected Archaea dying as a result, or unable to pass this trait along. But eventually, as evidenced by the fact that we have blue-green algae and plants today, it eventually stuck. An incredible origin story, to be sure, but has the process or chemistry of photosynthesis changed much over the past 2 billion years? No, not really. In fact, when this function evolved it was not only very complicated but also very inefficient. Take an analogy of photovoltaic solar energy. Normal commercial solar panels are about 20% efficient, meaning that they convert 20% of the photons that hit their surface

Figure 8.3 ***Folded and Fractured Sedimentary Rocks*** The geological record is not always easy to read.

into electricity. Efficiency is even better in laboratory conditions—up to 50% conversion. Compare those numbers to plant photosynthesis. On average, a plant converts only 1% of the photons that hit its surface into energy. And it is even worse for most of our crop plants—about 0.5%. It seems like 2 billion years would have given a lot of time for plants to do a better job, but in fact, sometimes evolution does things "just good enough." The "if it's not broken, don't fix it" model of biochemistry!

Evolution cannot be completely separated from the propagation, or survival of a species. In the case of photosynthesis given above, the evolutionary pathway that led to this particular biochemical process sort of stuck (except for the interesting mutation that occurred tens of millions of years ago that resulted in the C4 photosynthetic pathway, discussed in Chapter 7. But of course, photosynthesis has been adopted by a countless range of different plants over time, ranging from the lowly blue-green algae to the wispy ginkgo. In both of these cases, not only did the process of photosynthesis work, but the plants themselves were able to propagate and survive in their environments—in the case of algae, for billions of years, and ginkgo, hundreds of millions (www.kew.org/read-and-watch/ginkgo-biloba-maidenhair-tree-kew-gardens).

The pathway of evolution for modern humans (www.nature.com/articles/s41586-021-04275-8?WT.ec_id=NATURE-20220127&utm_source=nature_etoc&utm_medium=email&utm_campaign=20220127&sap-outbound-id=AE1E07D86F85AFF6677BE34763952B81970FA6A9) has been marked by several broken threads, and long periods of stagnation followed by rapid fits of evolution. Species have to be able to survive and propagate through those stagnant times, but this survival can be precarious. For example, the human forebearer Homo Erectus survived for about 2 million years—a solid record gave the advent of the northern hemisphere Ice Ages and consequent host of environmental shifts that occurred between 2.1 and 0.1 million years ago. Homo sapiens, on the other hand, have only been around for about 230,000, and we were almost wiped out during the cooling period soon before our last Ice Age, where our entire species was down to between 600 and 10,000 breeding pairs due to droughts, floods, and a huge volcanic eruption about 70,000 years ago (https://www.npr.org/sections/krulwich/2012/10/22/163397584/how-human-beings-almost-vanished-from-earth-in-70-000-b-) These numbers would have placed us on the Endangered Species list! But survive, and propagate, we did, reaching well over 2 billion breeding pairs today.

Early hominid mutations and the "back-to-Africa" concept

All humans and nonhuman apes have a specific genetic mutation that causes them to have much higher levels of uric acid in their systems than other organisms. We know uric acid as the namesake for one of our waste products, urine. And indeed, uric acid itself is a waste product of another chemical in our bodies. High levels of uric acid in the system cause an entire cascade of diseases, including gout, hypertension, diabetes, and obesity. Why were we so unlucky with our high uric acid levels, when actually this was the product of a welcome mutation among apes living in Europe between 17 and 12 million years ago, the Middle Miocene period? It all comes down to fat and sugar.

The Middle Miocene was a time of global cooling, a progression that began earlier in the Cenozoic Era and continues today (as thoroughly described in Chapter 7). This cooling was related to changes in continental configuration, increased removal of the greenhouse gas carbon dioxide from the atmosphere from weathering reactions, and a general slowdown of the tectonic engine that would typically return that carbon dioxide back to the atmosphere. One early result of this global cooling was a drop in sea level, which resulted in a land bridge opening up between Africa and the Middle East. Early apes, giraffes, camels, elephants, and all types of African fauna expanded across this land bridge. In particular, apes spread throughout Europe. But formerly balmy Europe was just starting to become the Europe of chilly winters that we are familiar with today. These early hominid settlers were increasingly faced with long intervals of lean conditions during winters in the now colder European climate, likely placing their very survival at risk. Then, evolution came for the win.

A genetic mutation during this time resulted in a change in the synthesis of uricase, an enzyme that normally keeps levels of toxic uric acid in check and is tightly coupled to sugar metabolism. It has been conjectured that this mutation, which results in relatively high levels of uric acid in our systems and those of all great apes, provided a survival advantage for the harsher Middle Miocene winters by increasing fat production. This fat store could be metabolized by early hominids during winter when very few other food sources were available. Seems like a far-out conjecture, but there is significant paleoanthropological evidence to support it, painted in the genes of African hominids and their descendants.

The so-called "Back-to-Africa" concept is based on the fact that great apes, and we humans who descended from them, exhibit this same mutation—even those who always dwelled in Africa (https://onlinelibrary.wiley.com/doi/abs/10.1002/evan.20266?casa_token=liLhiwJ2OywAAAAA:QQdi-L7g0pGi8hBl6akrAPjLNUWOwDS4oGM0R7hwd6wyh50nbJ9NVQpv5XmgiyYywAl79ShE3egVAGE). In other words, apes gained this mutation while in Europe and, likely in response to progressively colder and harsher conditions in Europe, brought it back to Africa, where it propagated. The retention of this mutation not only makes no sense in the tropical African environment that doesn't experience the same type of seasonal cooling and resultant starvation, but it could, and often is, harmful. Take for example modern humans, for which high uric acid content can lead to an increased risk of kidney disease, obesity, and diabetes. Before the 19th century, this likely was not a widespread issue because refined sugar (recall that uricase interacts with sugar) was not widely used and thus sugar itself was only a minor part of the typical diet. But the globalization of the sugar market and huge increases in the use of sugar in households and in packaged foods have caused this so-called "thrifty gene," meaning one essential for species' survival in lean times, but harmful to human health in times of plenty, to become a huge health issue now (https://onlinelibrary.wiley.com/doi/full/10.1111/acer.14655). Interestingly, during the similar Middle Miocene time frame, another mutation likely occurred in response to hominid consumption of fallen, often fermented, fruits. This mutation also similarly increased fat storage but resulted in biochemical conditions that increase the risk of alcoholism—another human health issue.

Neanderthals iced out of existence

As noted earlier, our most recent hominid coinhabitant of the planet was Homo Neanderthalensis Fig. 8.4, named after the Neander valley where fossils of this organism were first identified. Neanderthals were not really our ancestors—we didn't descend from them but we certainly bred with them and do share a common ancestor. Our two species likely diverged from each other more than 500,000 years ago, well into the Pleistocene period of extreme glacial/interglacial climate fluctuations. Whereas *Homo sapiens* evolved in Africa, Neanderthals evolved in Europe and Asia, and thus their physiology was shaped by the climatic variations around them and by the landscapes and food that they encountered. Their short, stocky build helped against the cold conditions they evolved in—less lanky extremities mean relatively less heat loss. Their great strength was suited to ambushing and taking down prey animals. They were intelligent, perhaps as intelligent as homo sapiens (remember that the next time you use the term "neanderthal" to denigrate someone's intelligence!) and they shaped stone tools, which they used to attack their prey at close distances, a risky proposition when facing off against a mammoth. They created art Fig. 8.5, would bury their dead, and, based on jaw and ear structures similar to humans,

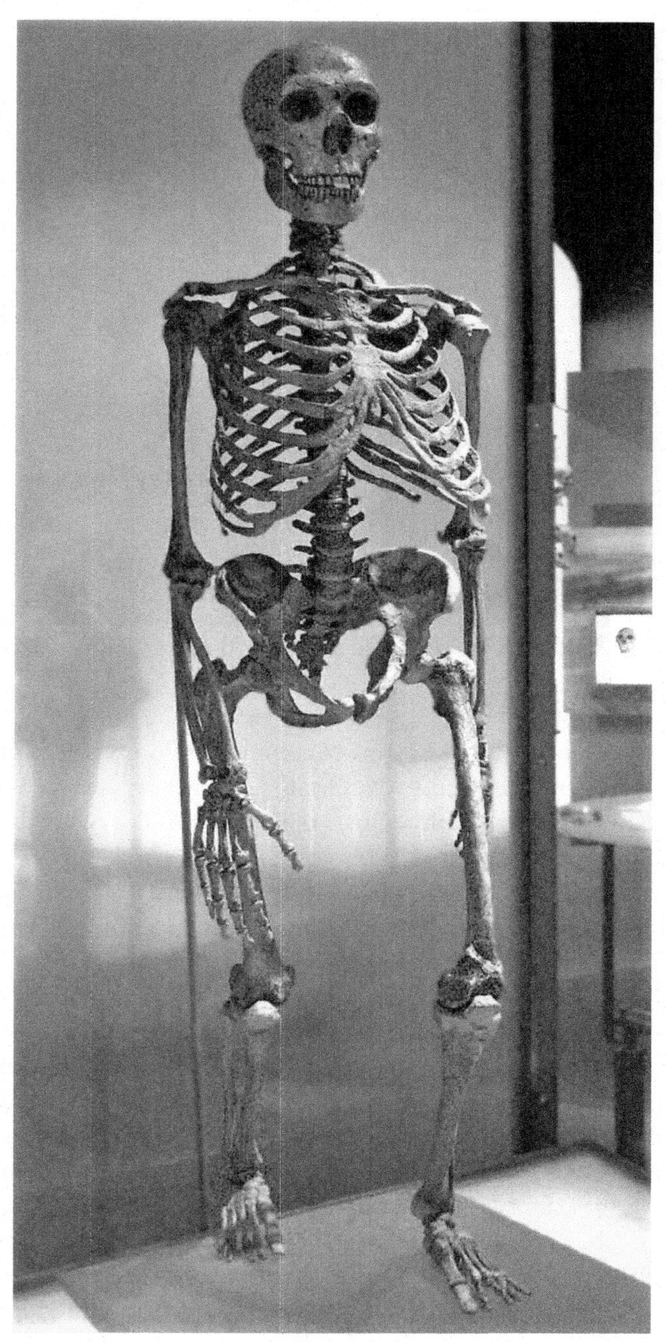

Figure 8.4 ***Reconstruction of a Neanderthal Skeleton*** Our cousins the Neanderthal were stronger than us and likely as intelligent and creative. Their mass required huge caloric intakes to maintain, and they eventually became extinct, with remnants of their DNA persisted in most of us.

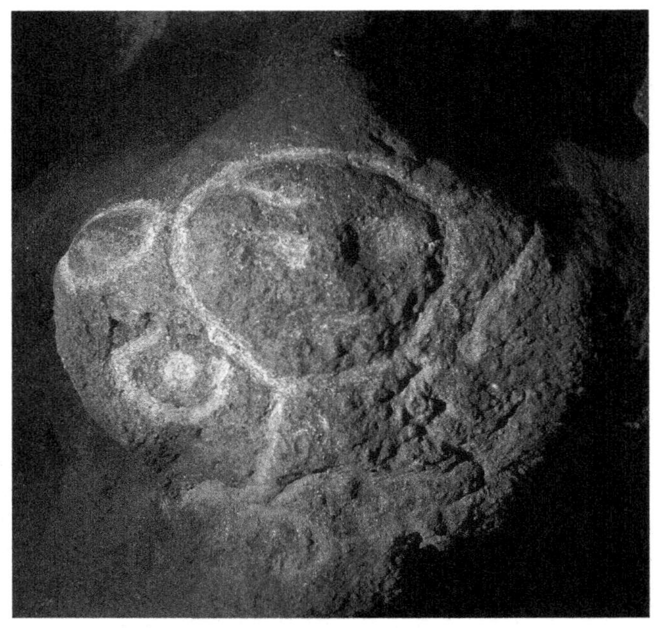

Figure 8.5 ***Neanderthal-Era Petroglyph*** Neanderthals produced art, including this rock carving.

they might have had advanced forms of communication (www.nhm.ac.uk/discover/who-were-the-neanderthals.html). And they survived through harsh ice age conditions, with animals as their main diet, with the earliest known fossil evidence from 430,000 years ago and the last fossil evidence from 40,000 years ago, after which they vanished. But after surviving through harsh conditions, and with similar mental resources as humans, what killed them off? Was it clashes with humans, climate change, or something else?

Two extant theories to explain the Neanderthal demise revolve around us humans being the culprits, and climate change. *Homo sapiens* coexisted in time with Neanderthals, but not necessarily in space. In fact, the first evidence of homo sapiens appearance in Europe was 45,000 years ago, and the last sighting of Neanderthals was only 5000 years later. One theory notes that Neanderthals already were in a population decline by the time of human arrival, and highlights this period of overlap in space and time could have been marked by either violent clashes between the species or diseases spread from humans to Neanderthals. There is evidence of the former, and there is certainly enough precedent for diseases introduced from one group into an immunologically "naïve" population without protections, resulting in widespread disease and death. For example, the colonizing Europeans brought smallpox, to which they had already developed some resistance, to indigenous populations in the Americas, resulting in the death of up to 90% of the indigenous populations where the contact occurred, in some areas within 50 years. So, the 5000 years of potential contact between humans and Neanderthals might have been enough to lead to their extinction.

Another theory places the blame on climate change (www.smithsonianmag.com/smart-news/modern-humans-didnt-kill-neanderthals-weather-did-180970167/, www.pnas.org/content/115/37/9116), based on evidence from cave deposits called stalagmites. Stalagmites form from water dripping through the soil into an underground cavity, or cave. Stalagmites grow layer by layer, year after year, and thus researchers have capitalized on this characteristic to use the stalagmite layers much like climate scientist use layers in ice cores or botanists uses tree rings to explore past environmental conditions. Researchers recovered cores from stalagmite deposits found in two caves in Romania to estimate local temperatures and rainfall conditions and found two intervals of extreme cold and dry conditions, one occurring 44,000 years ago and lasting for a 1000 years and another 600-year-long event shortly afterward at 40,800 years ago. Average temperatures during these intervals were below freezing, and thus the ground developed permafrost. Given these harsh conditions, it is likely that the large animals that Neanderthals hunted and ate either died or migrated away, resulting in the starvation and death of Neanderthals in two frigid waves.

It remains puzzling that they could make it through other climatic variations but not this one, so perhaps humans aren't completely off the hook for their extinctions. It is clear though that even with our brief period of coexistence, Homo Neanderthalensis and Home Sapiens did indeed interact, and interbreed (www.bbc.com/future/article/20210112-heres-what-sex-with-neanderthals-was-like). Modern humans owe about 2% of our genetic makeup to Neanderthals–and the author much more than that, at least based on genetic testing results that indicate I am 94% more Neanderthal than the average person! So perhaps humans simply absorbed the already declining Neanderthals into our population groups. How could two species interbreed and have sexually viable offspring? First, the definition of species is not as rigid as you might think, and it becomes murkier when using fossils to make the call. The DNA recovered from Neanderthal fossil remains does indicate that we were able to interbreed and produce sexually viable offspring, and even more than that—these offspring gain some valuable traits from their Neanderthal side of the family that protected against various diseases, making the now dominate Homo Sapiens more adaptable to the new environment and climate that they soon began to dominate.

And an additional discovery in 2022 from a cave called Grotte Mandrin, in the Rhône Valley in southern France, might further muddy the Homo Sapien-Homo Neanderthalis coexistence waters. A child's molar tooth, along with hundreds of stone tools dating back about 54,000 years ago, discovered in the layers of a long-occupied cave suggests that humans lived in Europe about 10,000 years earlier than thought, and thus had an overlap of over 15,000 years with Neanderthals. Critically, the human tooth was found intermixed between layers of Neanderthal remains, indicating coexistence, and potentially even peaceful coexistence, between the two human groups in the region. According to this research, "These findings challenge the narrative that the arrival of *Homo sapiens* in Europe triggered the extinction of Neanderthals, who lived in Europe and parts of Asia for about 300,000 years before disappearing." Science continues on, and we likely have not heard the last of the human-Neanderthal drama! https://edition.cnn.com/2022/02/09/europe/tooth-human-neanderthal-france-cave-scn/index.html.

The global march of humans

Humans survived through several glacial cycles, and although our survival was at times precarious, we obviously have a number of characteristics that proved successful for survival, for now. But we didn't just survive, we thrived and spread, first out of Africa to Europe and through the Middle East to Asia, then to the Pacific and Australia and the Americas. This diaspora occurred through significantly changing environmental conditions on land as the climate shifted. Given our roughly 200,000-year tenure on the planet, our evolution saw the Earth swing from a moderately warm interval to a deeply cold one that was almost the death of our species, followed by a protected warm interval, then a slow climb down into the depths of the last Ice Age around 20,000 years ago before the thaw and relative stability of our climate over the past ~10,000 years—the Holocene. This most recent period provided the climate stability to encourage communities to settle, to build more complex community structures than were needed in a more transient migratory community, and begin growing crops in earnest. Throughout this global human diaspora, humans began the process of developing racial heterogeneity driven by genetic regulation to fit the environmental conditions present (www.nytimes.com/2013/02/15/science/studying-recent-human-evolution-at-the-genetic-level.html). In this way, the European cluster of *Homo sapiens*, through genetic regulation, lost some of their original melanin so that their skin could synthesize enough vitamin D in the weaker higher latitude sunlight to maintain health. The characteristic physical features of East Asians might have been a "package deal" that emerged 35,000 from a genetic mutation that included more sweat glands for better thermoregulation in hot and humid environments—all the rest came along for the genetic ride. In both of these cases, it is an adaptation to a climate that drove racial distinctions, revealing the power of genetic variations for a species' survival in a changing environment.

Perhaps the biggest threat to humans came during the glacial period that started around 195,000 years ago, soon after our evolution as a distinct species. This brutal cold interval was also extremely dry, as all glacial intervals are, leaving much of sub-Saharan Africa inhospitable. Human population numbers plummeted and drove the surviving population into a small area along the South African coast. This glacial period lasted for about 70,000 years, and humans were effectively pinned to the coast, likely living in coastal caves that were exposed in marine terraces. But one characteristic of the glacial period was perhaps our savior—low sea levels. During glacial periods, a significant amount of water is locked up as continental ice, and sea levels drop substantially. During these glacial periods, sea levels can be as much as 120 m lower than during today's warm interglacial conditions. Not only does the sea level fall, but a vast amount of what was formerly seabed is exposed as the sea retreats. In low relief coastlines, such as on the USA southeast coast, the shoreline was over 100 km farther offshore during glacial times as compared to today. Indeed, remains of early settlements or structures are found submerged all over the world, stranded underwater by postglacial sea level rise. Whether these are the origin of the Atlantis myth or not, they are evidence that humans actively used this new landscape for habitation and food gathering. This was certainly true for the South African coast, where our early ancestors had the safety of coastal terrace caves and the bounty of the sea, and the now extended coast habitat afforded by the retreat of the ocean, effectively at their doorsteps (www.scientificamerican.com/article/when-the-sea-saved-humanity-2012-12-07/). Humans survived through this period, and once the glacial period was over about 123,000 years ago, emerged from their southern caves and began the expansion throughout Africa and then into Europe about

45,000 years ago. All human beings alive today are descendants of that relatively small band of survivors who found one of the few places in sub-Saharan Africa to weather the climate storm.

The timeline for human expansion into Europe, Asia, Australia, and finally the Americas is well documented and likely occurred in fits and starts depending on environmental and climatic changes and technological developments, such as housing, clothing, and tool-building skills. But the real story of humans as civilization builders begins 10,000 years ago, at the start of the Holocene. At this point, the postglacial climate had stabilized, humans were a global presence, and the kernels of city-states and settled agriculture were sprouting. This particular aspect of stabilized populations is the focus here because typically, if climate changes and thus ecosystems shift, more mobile populations can just follow their preferred ecosystem, migrating with the climate zone. But settlement requires investment, whether in the agricultural systems with irrigation and terracing or in other physical infrastructures such as buildings and roads. The costs of climate change are much greater for fixed species, be they coral reefs impacted by rising ocean temperatures and ocean acidification, or human settlements (www.bloomsbury.com/us/climate-change-in-human-history-9781472598523/). Consider for example the vast infrastructure of US coastal cities, like Miami and New York, and what it will mean for these cities when the sea level rises several feet by the end of this century.

The relative climate stability seen during the Holocene is, indeed, only relative compared to the seesaw swings of glacial and interglacial periods, which were driven by orbital variations. The more subtle climate changes during the Holocene were largely related to volcanic eruptions, changes in solar activity, and ocean-atmosphere oscillations. Massive volcanic eruptions emit significant sulfate aerosols into the stratosphere, reflecting incoming solar radiation back out to space and cooling the plane for several years. Solar activity and sunspots vary cyclically, and can slightly impact incoming solar radiation. And internal variations in ocean circulation and atmospheric pressure cause redistribution of heat and precipitation, as exhibited by the El Nino Southern Oscillation system. But on longer timescales, these factors are typically secondary climate drivers behind orbital variations and greenhouse gasses, but for the shorter window of the Holocene, these major drivers are not key variables.

There are exceptions to this, with the Industrial Revolution ushering in significant and ongoing increases in carbon emissions to the atmosphere. Prior to the Industrial Revolution, various changes in land use through time likely altered carbon dynamics somewhat. Some examples are flooding agriculture likely practiced for thousands of years driving increased methane emissions, and subsistence forest clearing by fire practiced by many indigenous populations releasing carbon into the atmosphere. Finally, the driving factors for some climatic events during the Holocene are not well understood. An example of this is the African Humid Period (https://en.wikipedia.org/wiki/African_humid_period) that occurred in the early to middle Holocene, which resulted in the greening of the Sahara Desert and northern Africa and saw the birth of Pharaonic culture in Egypt. The list of factors driving this is long, including orbital variations, changes in polar circulation, greenhouse gas variations, warming of the Mediterranean Sea, changes in Earth's geomagnetic field, and many others. But clearly, even the climate change that occurred in the "stable" Holocene impacted people, and societies.

Climate change and societal collapse

Like any good postapocalyptic story in media, the concept of societal collapse has always had a special cache. It is perhaps the most obvious question that swirls around archaeological conferences—what made Society X collapse? And indeed, archaeologists have used an array of physical remains, including excavations of building sites, examination of tombs, and the historical record to try to understand why societies failed. Some societal collapses seemed so sudden, such as the rapid abandonment of Mayan cities, as to invoke a sudden shock or disaster. Others, such as the slow decline of the Roman Empire, were extremely well-documented and were related to a host of internal and external pressures. Climate change has been invoked as one of the drivers of many societal collapses, and here we explore four collapse examples: The Akkadian Empire of Syria and Egypt, the final blow to the Roman Empire, the aforementioned Mayan Empire of Central America, and the Mound Building Indians of pre-Columbian North America. Each illustrates the interconnectedness of climate change, population, natural resources, and resilience to societal health, and collapse.

Even during the very early part of the Holocene, the so-called "Age of the Humans," we seem to have already been doing a lot to influence global climate. Many studies show the influence of even subsistence-scale human agricultural and hunting practices on the local environment and various geochemical cycles (https://link.springer.com/article/10.1007/s10933-009-9372-1). But an early provocative paper by the paleoclimatologist Bill

Ruddiman was perhaps the first to argue that human agriculture changed the global methane cycle—recall that methane is an extremely powerful greenhouse gas—and thus warmed and somewhat stabilized the global climate (www.sciencedirect.com/science/article/abs/pii/S0277379101000671 www.sciencedirect.com/science/article/abs/pii/S0277379108000760). Further evidence has mounted that the human footprint on the global environment over the past 12,000 years (www.pnas.org/content/118/17/e2023483118), albeit MUCH less strong than since the Industrial Revolution, was nevertheless influential in shaping the planet. For example, even at the beginning of the Holocene more than 90% of temperate and tropical woodlands were inhabited by humans, and thus the characterization of certain landscapes or ecosystems natural, intact, or wild landscapes is elusive, so long as one disconnects humans from that characterization (www.pnas.org/content/118/17/e2023483118). The elusive "Noble Savage," living in complete harmony with nature and leaving nary a footprint behind, is in a modern context a term that is colonialist, patronizing, and offensive. All organisms live in and shape their environment, from the parrot fish who chews through immense volumes of coral every day and produces the fine white sand of tropical beaches to the microbes that layer themselves into complex biofilms and shape the geochemical exchange of elements from sediments to overlying water, organisms shape their environments. Humans simply have done so more profoundly, and have developed these environment-shaping actions much more quickly, than other organisms.

Akkadia and the case of the missing earthworms

The mighty Akkadian Empire dominated Syria and Iraq between 2400 and 2200 BCE. With a rich history of preserved stone craftwork Fig. 8.6 and language, and the technology to smelt metals such as copper and lead, this long-lived culture was arguable the first "Empire" in recorded history (insert figure here). Akkadia was rich in grain crops and livestock, and became somewhat of a breadbasket for other regions, with significant export of grain products from Akkadia and import of most everything else, including metal ore. As with any empire spanning 2000 years, Akkadia suffered through various geopolitical shifts, but for the most part, the culture remained resilient to many internal and internal challenges, until the global climate shifted about 4200 years ago.

The so-called "4.2 kyear aridification event" is recorded in some paleoclimate records, but not all (www.science.org/content/article/massive-drought-or-myth-scientists-spar-over-ancient-climate-event-behind-our-new).

Figure 8.6 ***Bronze Head of an Akkadian Ruler*** Bronze head of an Akkadian ruler, discovered in Nineveh in 1931, presumably depicting either Sargon or, more probably, Sargon's grandson Naram-Sin.

Additionally, the search for the aridification event began first with the apparent contemporaneous collapse of empires like Akkadia and others, such as the Liangzhu culture in the Yangtze River region and the Indus Valley Civilization. The chief proponent of the climate-collapse connection for the Akkadians was archeologist Harvey Weiss who, upon excavating a site in northeast Syria with colleagues, found a sharp layer of wind-blown silt without any evidence at all of reworking by worms—in other words, a profoundly desolate interval of desertification (www.nature.com/articles/d41586-022-00157-9?WT.ec_id=NATURE-20220127&utm_source=nature_etoc&utm_medium=email&utm_campaign=20220127&sap-outbound-id=AE1E07D86F85AFF6677BE34763952B81970FA6A9). Searching in other regions beyond ancient Mesopotamia, Weiss found civilizations disrupted by aridification and strongly argued for a global climatic shift at that time.

This assertion, that climate must be to blame, drove many paleoclimatologists to look for confirming evidence, and many found it, including in cave deposits, which are particularly sensitive recorders of changes in rainfall. There was even a demarcation placed in the geologic timescale to indicate this "global" shift in climate. The only problem is that it might be all wrong. Upon further review, many other paleoclimatologists found either no evidence of a significant global or even regional climate shift—it doesn't makes an appearance in the typically robust Greenland Ice Core record—or evidence for a series of dry/wet/dry events over the span of a few centuries. As with the next examples of collapse presented here, the current understanding of climatic controls on the health of civilizations is simply not adequate to confirm a single cause of the collapse.

The final blow to the Roman Empire during the "worst year to be alive"

Countless responses can arise when the question "historically, what is the worst year to have been alive?" is posed. 2020 and the birth of COVID, which has killed 10s millions globally? 1942 and the ratcheting up of the Holocaust, the largest genocide in human history? 1918 and the start of the Spanish Flu, which eventually killed 50–100 million people, many of them young? 1860 and the fracturing of the United States? 1349 and the start of the Bubonic Plague, which killed almost 1/3 of the European and Asian population? All of these are appropriate responses, and are heavily influenced by the timeframe and place that is being considered. But, for many historians, 1 year, and 1 year only, qualifies as the worst year to be alive—536 AD (www.science.org/content/article/why-536-was-worst-year-be-alive).
Perhaps that year doesn't ring many bells, as it is buried in the deep past of our recorded history, but by all accounts, it was a terrible one (as were the slightly less terrible but still bad several years that followed). 536 AD was marked by several massive volcanic eruptions, widespread crop failure and famine, freezing temperatures, the rapid spread of diseases, death, and the final fracturing of the Roman Empire.

Historical records revealed the social and environmental shocks in 536 AD. The Byzantine historian Procopius documented a mysterious 18-month-long fog that plunged much of Europe Asia, and the Middle East into a twilight that lasted for 18 months. The summer of 536 was remarkably cold, with average temperatures dropping by about 2°C followed by a decade of similarly chilly weather. Snow fell in China … in the summer. Crops failed, and people starved. Then, as if things couldn't get any worse, a massive plague broke out in the Roman port city of Pelusium. Known as the "Justinian Plague" this disease swept through the East Roman Empire and killed between 1/3 and ½ of the population, hastening the eventual demise of the Roman Empire. Historians thus have a pretty solid justification to say 536 (and the years that followed) was a terrible year to be alive, but what drove this collapse?

A glacial ice core recovered from the Swiss Alps has helped to scientifically solidify 536 in historical infamy. Ice cores can provide annual and even subannual reconstructions of past climate and environmental conditions. This particular ice core revealed a layer of ash that can be traced to a massive volcanic eruption in Iceland at that time. Iceland is a tectonic "hot spot" and frequently exhibits large eruptions, but this one clearly had a regional, and even global impact on human societies. The eruption itself didn't harm those who lived downwind in Europe, the Middle East, and Asia, but the ash that was emitted from the eruption, that same ash that was documented in the Alpine ice core, certainly did. We have many contemporary records of the cooling effects of large volcanic eruptions. The 1991 eruption of Mt. Pinatubo in the Philippines, for example, cooled northern hemisphere temperatures by as much as 0.5°C for 2 years by injecting millions of tons of fine ash particles and sulfate aerosols into the stratosphere, where they lingered for years and acted to bounce incoming solar radiation back to space before it could warm the planet. This volcanic cooling mechanism is tied to many of the Mass Extinction events documented in Chapter 6, although one-off eruptions like Mt. Pinatubo and the Icelandic eruption in 536 have only a transitory impact on climate. But sometimes transitory is enough to be the death knell of an Empire already in decline. A combination of failed crops and famine meant mass dislocation of rural populations to cities. There, they encountered the main vectors of plague, rats and the fleas that thrive on them, who were also fleeing their own rat famine because of a lack of wild foods. This concentration of people and a disease vector might very well have sparked the Justinian Plague. In this way, climate change, even temporarily, has a significant ripple effect on human populations.

The Alpine ice core didn't just reflect this disastrous history, but also, decades and centuries later the redevelopment of industries signaled some kind of recovery. In particular, the cores revealed layers of the element lead, which was a byproduct of silver mining and was swept up onto the ice through the emission of smelting and mining particles (https://agupubs.onlinelibrary.wiley.com/doi/full/10.1002/2017GH000064). These peaks signal a return of economic activity, as the silver was mined and smelted to make coins for commerce. Interestingly, this particular ice core also revealed the impacts of the subsequent Black Death epidemic of Bubonic Plague that decimated Europe and Asia and likely ushered in the birth of the Middle Class–through the death of a significant proportion of the serf class. An absolute disappearance of lead in the ice core at that time indicated a similar abandonment of mining and economic industries related to a global pandemic, but this one apparently not triggered by a climatic event.

The Mayan Empire—too many shamans, not enough scientists

Tikal is perhaps the most hauntingly beautiful ruin that I have ever seen. Partially excavated and featuring all of the major characteristics of the Classical Mayan Period, the site is tucked into the forest of central Guatemala on the Yucatan Peninsula. Standing in the central plaza, with a ball court on one side, pyramids large Fig. 8.7 and small on both ends, stelae (carved columns of unknown function seen throughout the Mayan Empire) on the other side, and howler monkeys serenading in the treetops, Tikal seems still alive with the spirit of the Mayans who would have been wandering the grounds over a 1000 year earlier. But here is the mystery that has fascinated me for so long—these metropolizes, symbolizing an incredible amount of work to build and maintain, were seemingly abandoned overnight. The forests, held at bay by the inhabitants, then marched right back to inhabit the structures that displaced them. The indigenous populations living near Tikal, and many of the other great cities of the Mayan Empire, knew little at all of the sites, or the people who populated them. It was as if the most powerful empire in Central America collapsed overnight. But researchers have begun unraveling the various environmental and social triggers for this "overnight disappearance," and increasingly consider the collapse of the Mayan Empire neither overnight nor truly a disappearance.

The Mayan Empire stretched across much of Central America, including southeastern Mexico and northern Central America. The Empire was not always a singular entity but included regional leaders, ebbs, and flows of

Figure 8.7 ***Temple One in Tikal*** A primary temple in the Mayan city of Tikal. Temple one stands at one end of a grand plaza, complete with a ball court and various stellae for recording important events.

conflict and rapprochement, and various regional leaders. It is broadly defined as starting as early as two to 3000 years ago, peaking during its Classical period from 250–950 AD when most of the major infrastructure, and famous pyramids, were built, then followed by a decline—the so-called "collapse" period. Our starting point, the city of Tikal, followed a similar history, including a dramatic near-total depopulation of its ~70,000 inhabitants after the Classical period. This was similar to many of the great southern Maya cities, with much of the population migrating northward, where the northern Maya Empire held on for several more centuries. The potential cause of this rapid decline includes internecine warfare, overpopulation, environmental degradation, and climate change. But evidence found from the lake bottom of Lake Chichancanab (www.nature.com/articles/375391a0), in the center of the Yucatan Peninsula, points to a combination of environmental degradation and climate change as critical factors.

Much as the decade-by-decade accumulation of ocean sediments can reveal past conditions of marine life and climate, lake sediment cores can serve the same function, and can more clearly reflect local conditions around the lake. Thus, lake cores can provide an excellent local-regional reconstruction of climate and the environment with little risk of mixing sources from other regions. In the case of Lake Chichancanab, it is perhaps an ideal situation to capture regional records. Many lakes have multiple stream inlets, outflows, and/or mix water in the deeper lake between groundwater and the lake itself. Sediment records from these kinds of lakes can be confusing or ambiguous to interpret, as changes in any one of these factors can influence lake levels and lake chemistry. Lake Chichancanab is a so-called "closed basin" lake, meaning that the only input is from direct precipitation and overland flow, and the only output is from evaporation. The bottom of the lake itself is clay so there is also no groundwater. In a nutshell, if the climate is wet, the lake will be full and relatively "fresh," and when the climate is dry the lake levels drop and the salts in the water become concentrated.

Researchers capitalized on this characteristic to reconstruct aridity in the lake basin. The key mineral that they used for geochemical reconstructions was gypsum, which as a salt crystallizes out of the water when the evaporation is high and settles to the bottom of the lake. It ends up that there was only one time in the past 7000-year record reflected in the Lake Chichancanab that had a significant gypsum layer—the very time interval when the Classic Maya civilization began rapidly crumbling. Using the isotope geochemistry locked up in the gypsum layer, these researchers estimate that rainfall dropped by a catastrophic 50% during this interval, and as much as 70% during particularly arid intervals (www.popsci.com/ancient-maya-drought-according-to-mineral/). There were actually two intervals of extremely dry conditions reflected in the record, one from about 750–850 AD at the height of the civilization, and another between 950 and 1050 AD, as political unrest and collapse rippled through the great Mayan cities.

The cause of these climate shifts seems to be related to swings in the tropical Jetstream, call the Intertropical Convergence Zone. This zone shifts north and south as a function of annual, decadal, and even longer oscillations in the average location of high-pressure zones, and with these shifts come significant changes in rainfall. The Maya Empire had little in the way of irrigation infrastructure, at least based on the archaeological record. So, no aqueducts or canals to transport water at a distance. The Yucatan Peninsula does have underground stores of freshwater, and many of these can be seen in the cenote cave systems that pepper the Peninsula to this day (interestingly, the circular distribution of the main Cenote systems in the Yucatan reflects the shock wave that resulted from the Chicxulub Meteor Impact that wiped out the Dinosaurs!). Even with these potential freshwater wells, no evidence remains of technologies to lift this water for surface agriculture. During arid times these Cenotes may have largely dried up anyway, and thus were not an option. The Maya had become concentrated in major cities, which required significant food systems to sustain—when much of that food system is based on rain-fed maize, and the rain stops, so goes the food.

Why the Maya weathered the first arid shock but not the second might have had less to do with climate and more with political stability, providing some resilience in the face of hardship. Perhaps soil stewardship was better, or the food system was more diverse, or distribution more effective. Regardless, the final Mayan collapse was likely spurred in no small part by climate change (www.science.org/doi/full/10.1126/science.aas9871). If the Mayans had employed more paleoclimatologists, who could have pointed to periodic swings in the aridity patterns and argued for more resilient agricultural systems, and fewer shamans whose solutions were sacrifices, the Mayans might have weathered the storm. But alas, the shamans tried, and failed, to coax water from the sky, and the people in the mighty Mayan cities drifted back to the jungle, leaving behind advanced systems of architecture, writing, and mathematics, to pursue subsistence-level farming, where many remain to this day.

The Mound Builders and the case of the missing corn

When current Americans think of the previous Indigenous inhabitants of the land, much of their perspectives are shaped by media representations, and stories of the forced removal of Native Americans from basically every landscape that the USA currently calls a country. So the portrayal is often of savage frontier clashes between Indians and white settlers, like the Little Big Horn and General Custer. Or overly romanticized accounts of Native Americans living as one with their natural surroundings. Or sad accounts of the Trail of Tears and the inhumane displacement of Indigenous populations. All of these portrayals are rooted in truths, the last one in particular, but a valuable question can be "what Indigenous societies existed before the arrival of European Colonists." The answers to this question are diverse and often depend on artifactual evidence for reconstructions.

Many people have heard about, or even visited, the cliff dwellings and rich ceramic history in the Mesa Verde and other areas of the US Southwest, remains of the ancestral Pueblo tribes termed Anasazi (www.nps.gov/meve/learn/historyculture/cliff_dwellings_home.htm). This cliff-dwelling society flourished from about 1200–1300 AD, and included a range of systems that we would equate with modern cities, including grain depots, cooking and storage/trade facilities and defense systems to protect against unwanted intruders. Then suddenly, these cities were abandoned. And presumably, the populations simply moved southward. The depopulation might have been from a combination of external resource pressures and climate change, but the relics that remain, largely unaltered from their time of formation owing to the dry desert air, reflect a fascinating window onto a brief and unique period of human habitation in the area.

Far less attention, however, has been paid to the Mound Building populations of the Mississippi and Ohio Valley regions. Perhaps that is because their main artifacts are literally large mounds and pyramids of stacked earth and turf, rising modestly above the midwestern landscape Fig. 8.8. But the Mound cities often housed a large population, likely numbering in the

Figure 8.8 ***A Platform Mound at Aztalan State Park in Aztalan, Wisconsin, USA*** The Mound building culture was spread throughout the Mississippi and Ohio River valleys in the Midwest, USA.

thousands on a seasonal basis at least and far outstripping the 100 or so individuals that inhabited the Cliff dwelling communities in the Southwest. Recent work on the history of the Mound Builders has uncovered a fascinating intertwined story of religion, strife, an agricultural revolution, and climate change that is reshaping our understanding of the growth and collapse of the Mound Builders.

The largest of the Mound Building settlements is Cahokia, just across the Mississippi River from modern-day St. Louis, Missouri. Indigenous groups of the Mississippian tribes began developing the Cahokia site around 800 AD during the so-called late Woodland period. Over the arc of a ~500-year history, Cahokia became the largest and likely central hub of a host of other "sister-cities" that were built in Illinois, Indiana, and variously around the upper Midwest USA. This network of mound cities, and various artifacts found within and around them, represent a long period of time in urban and societal development that was previously unknown among the pre-Columbian North American peoples. And then, after tens of cities, and hundreds of structures per city, were built, with the population of Cahokia likely exceeding tens of thousands of people, and with trade routes that spanned thousands of miles, the entire system came crashing down over a short interval of fewer than 100 years. What happened to the Mound Builders, and why did the collapse ripple through the entire society so quickly? As with all of these stories of collapse, the tale is complicated and still unfolding, but again, climate change and the lack of resilience likely played a role.

We have learned quite a bit about the Mound Builders from the artifacts that they left in and around their cities. These include chert, a hard glassy material that was imported from nearby Illinois and used for tools and weapon heads, copper, which was brought down from Michigan and smelted locally, and whelk shells, imported from the Gulf Coast and used for jewelry and ceremonial decoration. It also includes the metal lead in the form of galena, brought largely from Missouri and worked intensively on-site into tools, cosmetics, and ornaments—indeed, so much lead was worked that a clear layer of the metal is seen in lake sediment records from near the mounds (https://pubs.geoscienceworld.org/gsa/geology/article/47/12/1193/574378/Pre-Columbian-lead-pollution-from-Native-American), unrivaled in magnitude until the characteristic global peak in lead in sediments that came during the height of leaded gasoline use. The question remains unanswered whether the Mound Builders suffered similar negative health impacts from lead exposure as did people throughout the latter part of the last century, but given the widespread nature of lead working at that time, it would not be a surprise that at least some individuals suffered from early lead poisoning.

The discovery of this "lead layer" in lake sediments addresses one of the main recent additions to the paleoanthropological reconstruction of the Mound Building culture—namely, the importance that lake sediment records have played in reconstructing land use, agricultural development, and particularly climate change during this time. These sediment layers reflect the coevolution of social and cultural practices (such as lead working), agricultural developments, land use, and climate shift in this region. Together with the archaeological record, they reveal a society that started during a climatically auspicious time, built upon an agricultural revolution involving maize hybridization and widespread farming and consumption, and then collapsed as the summer rains that naturally irrigated the maize fields stopped arriving during a centuries-long drought.

The Midwest US is prone to longer-term climatic shifts as is most parts of the world. Although it experiences the short-term (i.e., several times per decade) shifts in temperature and precipitation related to the El Niño Southern Oscillation (ENSO) effect, it also seems to experience the sometimes century-long swings in a climate that occur related to conditions over the distant Pacific Ocean. These result in shifts from the Midwest receiving warm and moist air from the Gulf of Mexico during the growing summers, to more influence from the drier northern air coming from the north Pacific and sweeping over Canada. The Midwest is truly a crossroads of climate, even on a decadal scale, but when these patterns persist over decades or centuries, they can have profound impacts on agricultural stability, which in turn can cause societal stresses and warfare.

The artifacts and climate records from this region show that, as the Mounds were being developed and expanded, so too was a reliance on maize as the main food source. This shows up in the distinctive isotopic chemistry that maize brings with it. Corn uses a different photosynthetic mechanism (the C4 pathway discussed in Chapter 7) than, say, squashes and beans, and this difference shows up in the carbon isotopic chemistry in the lake sediment record and indeed in skeletal remains found in these sites (www.nature.com/articles/srep41628). We literally "are what we eat" in terms of carbon, and bones collected from burials near the Mounds sites reflect an increase in maize consumption over the traditional squash and beans, as does the carbon isotopic composition of contemporaneous lake sediment layers. This agricultural revolution allowed populations to store large amounts of grain and thus be more immune to seasonal food deficiencies, and maintain moderately stable caloric input throughout the year. It also made the population centers more vulnerable to potential looting from groups that didn't practice stabilized agriculture. indeed, as maize use and urban development proceeded through the interval of Mound cultural supremacy, so did the construction of various fortifications, including sometimes multiple rings of protective palisades that ringed the cities. Interestingly, most of these cities are along a river, presumably for easier transport of goods and people and access to water supplies, but apparently external threats did not come via canoe but rather overland, as palisades or other protective structures are not seen lining the riverbanks.

The zenith of the Mound building culture approached in 1350 AD or so, marked by huge settlements and vast permanent and migratory populations. Then, in a period of less than a century, the Mounds became depopulated in a staggered fashion throughout the Midwest, and the former inhabitants dispersed into the countryside and adopted new ways of life during the so-called "Vacant Quarter" of indigenous history when the formerly densely populated Mississippi and Missouri Valleys were abandoned. In a bit of a "broken record" moment, many factors contribute to societal collapse (www.nytimes.com/2021/04/24/science/cahokia-mounds-floods.html), and in this case, a severe and prolonged warm season drought was one of them, depriving the maize fields of their usual summer showers. Without irrigation systems, the crop yield likely declined to a level that could not support the higher population densities present in the cities. Even with the vast Mississippi and Ohio Rivers on their doorsteps, without significant efforts to move this water uphill to floodplain crop fields, the maize was doomed, and so was the Mound-building culture.

Summary

Our modern form of society, involving settlements, specialization, and organized trade, was forged in a period of climate stability. But homo sapiens weathered substantial natural climatic shifts, as did our hominid sisters and brothers the Neanderthals, and a long line of genetic ancestors. The range of behavioral climate adaptations displayed by later hominids included tool building, food preservation via fire, and thermal regulation via clothes and built housing. A similar range of genetic forms of adaptation, including skin melanin concentration and fat metabolism, were fully engaged over the past 200,000 years of human presence on the planet. And here we stand, perhaps the only organism to ever exist on this planet that is technically immune to climate change. We can cool and heat our surroundings, temperature control that, when added to clothing, means that humans can survive in the coldest, and the hottest, areas of the planet. Indeed, we can even survive in places that are completely hostile to all life, based on our stints in space, journeys to the Moon, and soon our incipient colonization of Mars.

All of this "climate control" comes with a host of caveats and conflicts. First, not everyone has access to cooling in the blistering hot summer. This access issue is true whether one considers a poor country, without an adequate stable supply of electricity for cooling, or a poor population within a country. This thermal injustice is nowhere as stark as the gleaming desert city of Dubai, where temperatures are sometimes so great, and electricity so plentiful for the wealthy, that outdoor walkways and shopping malls are cooled with air conditioning—not contained within a structure but just pumped out to the sidewalks and streets, soon to dissipate in the desert heat. Meanwhile, kilometers away, the working class of the city toil in unrelenting heat. And the conflict is clear—our ability to control the immediate temperature of our surroundings results in global changes to the temperature of the entire planet. The summer in Dubai would always hot, but the need for outside cooling systems is at least partially due to the warming effects of climate change. The temperature knob that we can dial in our house is also, at least in our current forms of producing electricity, in essence, largely also a temperature knob for the planet.

The final chapter will reflect on what that temperature will be set to in the future, and how global ecosystems, and the humans that share the planet with these ecosystems, will be impacted. But there is no question that the limits of human survivability on this planet are nowhere near tapped out, and that like our forebears, humans should, in theory, be able to handle the climatic changes around them—even those that they are actively causing. But might there be other genetic adaptations that are triggered by future climate change, or even the evolution of the next step in hominids that relies on technology? The long arc of hominid survival in the face of significant climate change bodes well for the future, but the future has the potential to be more difficult, and more unjust, if we don't embrace our own actions on climate change and other planetary assaults, so that our grandchildren are born into a less painful world. Modeling or forecasting how this will play out is open to far more uncertainty even than untangling the past geologic record of hominid-climate coevolution, and it is always prudent to heed the words of Nobel-winning atomic physicist Niels Bohr, who once said, "Prediction is difficult, especially about the future."

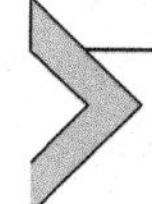

Did you know that?

We still don't know exactly what caused the extinction of the great megafauna of paleolithic North America

Imagine a time when mastodons and saber-tooth tigers roamed North America—these great megafaunas are iconic Ice Age mammals that hearken back to the days of the "cave men," but they actually persisted for far longer than homo sapiens and Neanderthals did. Mastodons may have been around for as long as 5 million years, living in North and Central America. They survived through the transition from the moderately warm Pliocene period to the Ice Ages of the Pleistocene. A species of mastodons, the iconic wooly mammoth with its thick coat of fur and long curving tusks, diverged from the mastodons about 700,000 years ago and was spread widely in the Arctic region, from Scandinavia through Siberia into North America. This beast was specially adapted to the cold conditions of the tundra regions, and could actually roam pretty freely across areas of the Arctic that were connected by ice sheets and land bridges during Ice Age intervals (when sea levels were low and the Bering Sea was actually a land bridge). The saber tooth tiger (formally the smilodon) was similarly iconic and appeared in North America starting about 2.5 million years ago, right when the Pleistocene was beginning and climatic conditions were becoming cooler. We have fantastic fossil records of all of these megafaunas, including from river deposits, peat bogs, directly from glacial ice, and even in the La Brea tar pits of Los Angeles, which hosts an impressive number of Smilodon fossils, animals that met their demise in the thick and noxious tars of the area.

Living and persisting through warm periods and Ice Ages, the Paleolithic megafauna seemed to be examples of resilient species that were well-adapted to the ever-changing climate of the region. So why is it that all of them, the mastodons and wooly mammoths and saber tooth tigers, after hundreds of thousands to millions of years of dominance in the Northern Hemisphere, suddenly vanished about 12,000 years ago? Some scientists conjecture that the extinction of the megafauna was due to climate change, the shift from the last Ice Age into our current Holocene warm period. But they had weathered that shift many times before, so why now? Other scientists point to the voracious appetite of humans and in particular the expansion of indigenous populations in North America, for the extinction. Ample evidence exists of the exploitation of the tusks and teeth of these animals for tools, and some scientists have conjectured that some sites in North America include evidence where indigenous tribes incited panic in herds of a mastodon, possibly brandishing fire to drive them over cliffs and to their deaths on the cliff floor below. Notably, though, the. most famous of such sites where Neanderthals supposedly used this practice has revealed that it was not likely to have occurred at this site, although the technique is not absolutely discounted as it was used by Native Americans on buffalo drives.

So, climate change, human predation, or the combination of the two factors are typical explanations for the rapid and absolute extinction of the classic Paleozoic megafauna. But what if the real cause was more otherworldly? An intriguing hypothesis was forwarded by Richard Firestone and colleagues in 2007 arguing that, based on evidence from an iridium-rich layer and fragments of tektites, shocked quartz, and nanodiamonds similar to observations from the dino-killing Chicxulub impact, maybe it was a ~4 km in diameter meteor that broke up in the atmosphere on entry and dealt the final blow to the mastodons, wooly mammoths, and saber tooth tigers. The layers of these impact evidence were relatively synchronous around the North Hemisphere, occurring about 12,500 years ago, and coincided with not only the last appearance of the megafauna, but the demise of the Clovis Man culture, indications of widespread wildfires at this time, and even the major climatic hiccup called the Younger Dryas interval, when the progressive retreat of ice sheets at the very end of the last Ice Age temporarily stopped, and indeed advanced again for a brief interval before finally retreating. The Younger Dryas impact hypothesis or Clovis comet hypothesis invokes a similar extinction driver as for the Chicxulub impact, but obviously, on a much more modest scale—impact, followed by ejecta-sparked wildfires, stratospheric injection of dust particles, resultant severe nuclear winter, and steep temporary decline in plant productivity, and then death and destruction. This hypothesis has some strong supporters, including those who feel that enough evidence has stacked up in its favor that it should be promoted to a theory (https://journals.sagepub.com/doi/10.1177/00368504211064272) to most others, who either dispute the evidence, dispute the timing, or dispute that one small impact caused that many problems. Science lives through the process of evidence, testing that evidence, posing hypotheses, and testing those hypotheses, and for now, at least, it looks like the impact theory for the extinction of the iconic Paleolithic megafauna will have to remain in the machinations of the scientific process.

CHAPTER NINE

Climate and life on future Earth

Introduction

The headlines can be downright apocalyptic. "Sydney is flooded, again, as climate crisis becomes new normal for Australia's most populous state"(https://edition.cnn.com/2022/07/04/australia/sydney-floods-damage-evacuation-climate-australia-intl-hnk/index.html) "More than 125 million people are under heat alerts across the US"(https://edition.cnn.com/2022/06/13/weather/excessive-heat-dome-forecast-wxn/index.html). "Climate change: wildfire risk has grown nearly everywhere—but we can still influence where and how fires strike" (https://theconversation.com/climate-change-wildfire-risk-has-grown-nearly-everywhere-but-we-can-still-influence-where-and-how-fires-strike-185465). "Massive coral bleaching hits Australia's Great Barrier Reef" (https://www.smithsonianmag.com/smart-news/mass-coral-bleaching-hits-australias-great-barrier-reef-180979823/). Those, among others, appeared in the press, in 2022 alone. Floods, fires, and extreme heat have become the new "climate abnormal" in many parts of the world. And, of course, it is getting very hot, with global temperature records broken almost every year Fig. 9.1.

Long predicted to bring havoc to global societies, and global ecosystems, climate change has settled in as a dominant factor in many weather disasters. That role is so large now that scientists can ascertain what increased likelihood of a given weather disaster is due to human-produced climate change, as well as how much additional damage costs were related to the climate change components of weather events. Because the weather has always impacted the planetary surface, basic dynamics like hurricanes and storms and heat waves were here long before human influence on climate and will continue long afterward. But many of the weather systems that cause damage to ecosystems and us are also responding to climate drivers, and thus the field of "climate attribution science" has been spawned to understand by what percentage, or with what increased likelihood, a given event was super-charged by climate change.

The speed at which the planetary systems are changing Fig. 9.2 from climate impacts has caught many off guard and is receiving important global

Climate Change and Life
ISBN: 978-0-12-822568-4
https://doi.org/10.1016/B978-0-12-822568-4.00002-X

Figure 9.1 ***Hottest Years on Record*** A graph of the hottest years since we have recorded global temperatures since 1880, showing that from 1880-2021, all of the top seven have been in the last seven years.

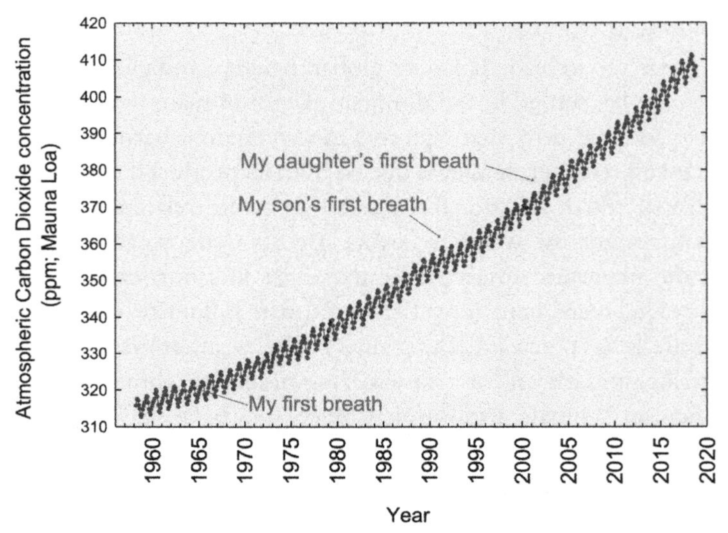

Figure 9.2 ***Concentrations of Carbon dioxide in the Atmosphere*** Scientists have continuously measured carbon dioxide in the atmosphere since 1958. Since that time, the rate of increase has been clear and increasing. Also included are the years of the author's birth, and that of his oldest son and youngest daughter, showing the sharp increase in this gas from one generation to the next, and within generations.

attention. Indeed, many positive actions are happening at local and global levels to get the key culprit, carbon emissions, under control and to figure out ways to actively "scrub" the excess carbon dioxide from the atmosphere. These efforts are made with the full awareness that our climate future is largely in our hands Fig. 9.3—we control the volume control on climate impacts, but of course, it is not as easy to change global energy systems as it is turning the volume control down on your computer. But we are increasingly confident about the potential range of future climate conditions until, say, 2100, which is a useful initial benchmark. We also have a good idea of how our own actions will have to change by 2050 to achieve some level of potential climate stability by the end of the century. Less certain, however, is how the various ecosystems on the planet will respond to this current rapid climate change. And unlike past mass extinctions, this sixth mass extinction is not just from climate change, but the myriad other ways that human activities have intentionally or inadvertently impacted global ecosystems. From deforestation to run-away urban development to contamination of water and air to loss of polar ice to direct exploitation of resources, the human footprint on the planet is not a light one.

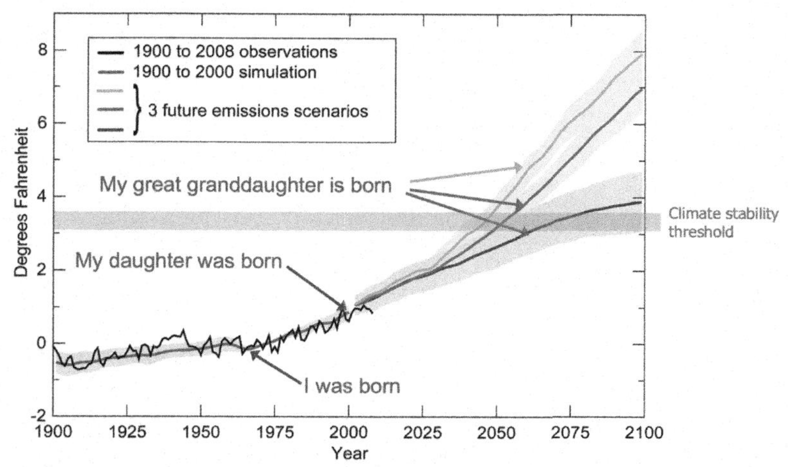

Figure 9.3 ***Future Climate Trajectories and the Estimated Stability Threshold of the Planet*** The trajectory ranges of future climate reflect humanities choices. Continue to emit high amounts of carbon will result in a future climate state that exceeds stability levels for our own descendants, whereas sharply limiting carbon emissions and working to extract carbon already in the atmosphere will bring levels closer to the stability threshold of the planet.

It is important not to become dispassionate about these real and present global dangers, but it is equally important to understand in a quantitative way how climate change and other disturbances are impacting earth systems that we all (the planetary biosphere) depend on and to predict as best we can the future arc of life on the planet. It is naïve to assume that we will use this sometimes quite a bleak portrait of the future to radically and rapidly change our current practices (although there are reasonable arguments for why we should), but it is equally naïve to think that we should just go on as usual and "hope things work out." Coming to terms with a changing planet should not breed Nihilism, but instead can be a critical tool for activism, providing benchmarks for action to protect against a future that we don't want our children or grandchildren to be born into.

To that end, this chapter will cover the current climate threats to marine and terrestrial ecosystems, what we can reasonably predict that the planet will look like in the agreed-upon climate time marker of 2100, and also provide the long view—how about 2300, or 2500? The incredible resilience of earth systems described in the earlier chapters of this book should give confidence that the planet, and life on it, will continue on. It will be different, as it always is when a climate-induced mass extinction drives a massive reshuffling of the global ecological deck. It might also be less nice for human beings, whose societies and technological advances were forged in a long period of exceptional climate stability, and were bolstered by a relatively untouched array of global resources, many of which are not unlimited. But there is some comfort, and humility, in thinking about the butterfly emerging from its cocoon and alighting on a flower 10000 years from now because certainly, that will be the case.

Threats

Much of this discussion of current threats and impacts will revolve around those caused by climate change, but it is sometimes impossible to untangle these direct impacts from the secondary effects of climate change. Take for example acid precipitation, or acid rain, as it is often referred to. Burning coal for electrical generation generates a significant amount of heat energy, but also a significant amount of nitrogen oxides and sulfur oxides that become acid volatiles. When released uncontrolled from the smokestacks of a coal-fired power plant, these acidic gasses dissolve into water vapor and become nitric acid and sulfuric acid. When the water vapor turns into a raindrop falling on a forest, the rain is much more acidic, with a pH that is often hundreds of times lower than normal rainfall. None of this is good for ecosystems downwind, as can be seen by the vast areas of forests and lakes that were devastated by acid deposition in the 1950's and 60's when few environmental controls were present in coal-fired power plants. The acid rainfall stripped valuable calcium from soils and stunted tree growth, and when that acidic water made its way down to lakes, it would alter lake chemistry and mobilize the element aluminum, which would literally precipitate out of streams and lakes into fish gills as they were trying to breathe, choking the fish. The legacy of acid deposition was vast areas of sterile and stunted forests, particularly in the areas downwind of primary coal-burning power plants. In North America, this was largely the forests and lakes of the northeast US and Canada—acidic rainfall produced by the coal plants in the Midwest didn't care much about the state or national boundaries, and many lawsuits were filed, and won, about this cross-border air pollution. Upon recognizing this environmental harm, the US Environmental Protection Agency in the 1970's and 80's mandated strict regulations as part of the Clean Air Act to place systems on power plants to capture and remove the acid volatiles before they escaped from smokestacks, and the forests and fish have largely recovered.

The above is a story of climate change—those same coal power plants that emitted acid volatiles also emitted vast quantities of carbon dioxide. But it also shows the secondary ecological impacts of climate change drivers. Many coal power plants still operate around the world, with some countries having stricter emission regulations releasing few air pollutants than in countries with more lax regulations. But regardless of how "good" the air is coming out of the smokestacks of a modern, technologically advanced fossil fuel power plant, it still pumps out about as much carbon dioxide as one of the older "dirty" plants that operated in the past and still do operate in many parts of the world. Carbon is carbon, whether it comes out of a coal power plant or a "cleaner" natural gas one. Whether it comes out of an old Ford Mustang or a new Toyota Prius hybrid. Yes, the carbon intensity per watt or mile is different, but the reality is that climate change is just a matter of greenhouse gas concentrations in the atmosphere, and nearly all of our uses of fossil fuels, even when they seem "green" like in a hybrid vehicle, still result in more carbon dioxide being emitted to the atmosphere every year than is removed by natural (and human-mitigated) processes.

Threats at sea

The vastness of the oceans and the ample resources available in them belies the fact that they, too, are vulnerable to the impacts of climate change. Intriguingly, on longer, geologic timescales (greater than 10,000 years) the oceans may be our carbon savior, as their capacity to absorb significant amounts of carbon on short and long time scales will help, eventually, to pull down human-produced carbon dioxide in the atmosphere. Almost half of all carbon emissions into the atmosphere are currently quickly absorbed by the ocean, while the other half is partially absorbed by the terrestrial ecosystem but largely simply accumulates in the atmosphere, causing anthropogenic climate change. This process of ocean absorption of carbon does not occur without its own negative side effects—much like the nitrogen and sulfur oxides from coal burning cause acidic volatiles and lowered pH of rainfall, so also the increased absorption of carbon into the ocean flips the buffering balance of the ocean water and reduces pH. On land, this spelled trouble for forests and fish and was relatively quickly brought under control once fully recognized. In the ocean, this acidification is also a very concerning negative side effect of carbon emissions but does not receive nearly the attention, or action, that it deserves. It is not just ocean acidification that is stressing marine ecosystems, but as discussed also in Chapter 8, marine heat waves, a direct product of climate change and warming, which may be having an even bigger impact on many systems.

Ocean warming

Impacts on marine ecosystems

The atmosphere and the ocean exchange many things, including water in the form of evaporation and precipitation, gasses like carbon dioxide and oxygen, and heat. As the atmosphere has been heating from climate change, so too has the surface ocean. Much like swimming in a lake in the middle of summer, where you will have a nice warm layer on top where your head is and yet still be a bit cold where your feet are, the ocean has similar thermal layers. The upper layer of the ocean is receiving much additional direct heating from the warming atmosphere, and much of the carbon dioxide, but that heat will slowly be transferred into deeper portions of the ocean through mixing. Most of the productive ecosystems in the ocean, those that play the largest roles in terms of biological diversity, exchanging oxygen to the atmosphere, and providing food for a hungry planet, are in the surface layer of the ocean. Marine heating causes changes in the migration pattern and habitat of many marine organisms, but the nonsessile ones, i.e., those that

Figure 9.4 ***Bleached Coral*** Marine heat waves can cause coral bleaching, a condition exacerbated by ocean acidification.

can move, do have some ability to migrate to areas that better suit their heat limits. Sessile organisms like clams and mussels and coral, do not enjoy that ability. They play out their life cycle mostly in one place, and if that place gets too hot for comfort, these organisms get stressed. In the case of coral, the increase in ocean heat waves causes mass bleaching events Fig. 9.4 and, when ocean acidification is added on as an additional ecological strain, many of these systems are buckling. Some resilience exists in even the sessile ecosystems, but there are limits to this. It is true that even sessile ecosystems can migrate, or at least their polyps can, but unlike a fish, there is little in the way of intentionality to a coral spawning event—the eggs and sperm are released into the water and go wherever the ocean currents take them.

For the nonsessile organisms of the surface ocean, there can be a mismatch between food source and comfort zone. Fish are consumers, and reef fish, for example, depend on healthy reef systems to survive. A major coral bleaching event causes a primary food and habitation source for most of the reef fish to disappear, and unlike the stressed coral polyps themselves, who can rerecruit their zooxanthellae symbionts when better conditions return in months to a year, the rest of the food web, which rely on healthy corals cannot. Thus, repeated bleaching events test the resilience limits of the entire coral reef ecosystem.

Other fish, largely those that live in the more biologically productive areas of ocean upwelling, may be directly affected by marine warming but their biggest threat is not climate-related, it is us-related. Overfishing and

Figure 9.5 ***Net Fishing Bycatch*** Undesired fish caught during net fishing are thrown back overboard.

unsustainable fishing practices have decimated fish populations, as well as those of many other organisms that are impacted by modern fisheries practices. So-called "bycatch" is all of the other organisms that some net fishing operations pull in Fig. 9.5, including dolphins, turtles, and living coral from the seafloor. All of these are thrown overboard, but obviously in a damaged state or typically dead. Other organisms like sharks are suffering massive declines because of the barbaric practice of shark finning, where the sharks are caught, their fins cut off for various medicinal products, and thrown back into the ocean to die. So, although ocean warming may impact the larger organisms of the ocean, it is hard to gauge due to the much more clear and present danger from marine overexploitation.

Taking the temperature of the sea

Interestingly, the extent of ocean warming is not just gauged by a series of floating and diving autonomous thermometers and by remote sensing from satellites, but also by a technique that in itself might pose risks to marine organisms. This might take some patience, but it all revolves around the "SOFAR" channel in the marine system. This is a layer in the ocean where sound is transmitted at great distances. And yes, physical oceanographers do have a sense of humor, picking first the acronym and then searching for words that fit it! This layer was discovered during the height of World War II to test better ways to detect enemy submarines at a distance. Scientists used a sensitive marine microphone lowered from a research ship out of one of the top marine science laboratories in the USA, Woods Hole, Massachusetts. About 900 miles away, another ship lowered a small explosive device at depth to make the sound signal, to see if the microphone would pick it up. Sure enough, the signal came through loud and clear, and a new way to transmit and receive messages over great distances was discovered. Well, discovered by humans, at least. Whales have been using this same "technology" for millions of years, transmitting their vocalizations across great distances in the ocean. These beautiful whale songs sound clear and sometimes mournful but were such an iconic example of animal language through song that they were recorded and sent up on a gold plate affixed to the interstellar probes Voyager 1 and 2, waiting for an opportunity to be discovered and replayed by other fellow inhabitants of our cosmos (both probes are now beyond our own solar system, safely carrying their earthly messages into deep space). But now, humans were sending out deafening explosions in the same sound channel that whales use to communicate and even share stories of sorts to distant pods. It would be akin to having a car alarm blaring when you were trying to enjoy music—in that same car.

Back to detecting ocean warming. The speed at which signals travel in the SOFAR channel is affected by the temperature of the water in that layer. Warmer temperatures equal faster sound travel and vice versa. Scientists in the 1990's began to capitalize on this physical characteristic of the ocean to use SOFAR as a large-scale marine thermometer, sending powerful sound waves through the channel and measuring how long the signals took to reach the microphone. The Acoustic Thermometry of Ocean Climate program was an international effort involving almost a dozen institutions in seven different countries. These partners all had sensitive microphones suspended in the SOFAR channel spread around the Pacific Ocean, each measuring how long it took for a strong signal emitted from an acoustic source (called a sonic gun) installed on the Pioneer Seamount off the central California coast. A sharp debate quickly ensued, though, from marine protection organizations. Sure, it is important to measure ocean warming and these experiments were a unique and powerful way to do so. But they also emitted an extremely loud signal at a very low frequency, and the impacts of this on marine biota were not adequately tested. Not only would it interfere with the low-frequency vocalizations used by whales, but near the source, it was so powerful that it might disorient, or even deafen, marine organisms like fish, dolphins, and sea turtles. Enough evidence linked these types of powerful submarine sonic signals to negative impacts on marine ecosystems that all such experiments are now carefully controlled to take into account the migration of marine organisms and the presence of sensitive species near the potential sound source.

Beyond the value of SOFAR for taking the ocean's temperature, research conducted in 2022 (https://phys.org/news/2022-03-oceans-significantly-underwater.html#:~:text=In%20addition%20to%20the%20notable,Arctic%20Ocean%2C%20Gulf%20of%20Mexico%2C) is examining how ocean warming will impact the general soundscape for marine organisms. These studies have predicted how the common vocalizations of the endangered North Atlantic right whale will change with various climate change scenarios, as their vocalizations are at the low-frequency range that will be greatly affected by increased temperature. This may be a benefit—they can "hear" farther—or a curse, if they assume a certain distance to a mate or food source but it is significantly farther away than expected.

Ocean warming and gas fluxes

Gasses exchange between the atmosphere and the ocean all the time, and this process is critical for marine biological productivity and for global carbon cycling in general. For example, the carbon dioxide that makes it way into the ocean, either directly by dissolving into the surface ocean from the atmosphere or indirectly via riverine carboonate ions, becomes a chief source of shell material for primary producers and is of course the fuel for photosynthesis by these organisms. Similarly, the oxygen produced by photosynthesis dissolves into the surrounding waters and is taken up either by marine respiration or leaks back into the atmosphere, making it possible for all of us to breathe. Gas exchange between the atmosphere and the ocean can be described by a simple gas law equation called Henry's Law, which defines the balance of a given gas molecule at a given temperature. Although that exchange equation is different for carbon dioxide than it is for, say oxygen, one factor is consistent—at colder ocean conditions, more gas can be absorbed out of the atmosphere by the water. That equation is symmetrical so that at warmer ocean temperatures the water absorbs less gas. It is also a reversible equation, and one can play around with this effect in their own kitchen, using the relative apparent carbonation of a warm and a cold soda—in this experiment, once opened, a warm soda loses its fizz much faster than a cold soda.

As surface ocean waters are warming from the transmission of atmospheric heat, their gas absorption properties have changed as predicted by Henry's Law (and which can be modeled in the future by applying this same gas law to various climate change scenarios). This has immediate implications for the two most important marine gases—carbon dioxide and oxygen. For carbon dioxide, this surface warming means that the relative balance of gas between the atmosphere and ocean is shifting, with the ocean less able to absorb the same fraction of carbon dioxide even as the atmospheric concentrations of the gas have increased from human activities. In other words, the ocean is increasingly less able to absorb the excess carbon dioxide, somewhat limiting its role as a carbon sink (https://news.wisc.edu/climate-change-reducing-oceans-carbon-dioxide-uptake/). This "positive feedback" loop is, as with most of those related to climate change, actually a very negative thing. More carbon dioxide in the atmosphere equals more warming equals increased ocean temperatures equals decreased ability to absorb carbon dioxide equals more carbon dioxide in the ocean ... rinse and repeat.

Ocean warming also decreases the diffusion of oxygen into the ocean, with negative impacts on many marine ecosystems. Many of these ecosystems are already under "oxygen stress" due to the inadvertent fertilization of coastal marine systems from agricultural runoff, causing eutrophication and low oxygen concentrations. This oxygen deprivation is now being compounded by a global decrease in ocean oxygen uptake from the atmosphere (https://www.scientificamerican.com/article/marine-oxygen-levels-are-the-next-great-casualty-of-climate-change/), potentially driving some ecosystems into a low, or no oxygen state, which spells death for any animals that live in those environments. Large areas of the ocean have lost up to 40% of their oxygen, a situation that will be compounded by ongoing climate change and could impact global livelihoods and nutrition losses.

Altered ocean circulation

Geologic evidence from deep-sea sediment cores has revealed many of the dynamics of ocean circulation and related shifts in ocean density driven by salinity and temperature. One critical circulation pattern in the Northern Hemisphere is the Atlantic Meridional Overturning Circulation or AMOC.

This pattern defines the exchange of water and heat over much of the Northern Hemisphere, including bringing tropical Gulf Stream waters into the otherwise frigid North Atlantic region. The strength of the AMOC is dependent on how much water can be made dense in the North Atlantic to be exported as North Atlantic Deep Water. Saltier and colder conditions in the North Atlantic speed up the AMOC and fresher and warmer waters slow it. A fast AMOC means warmer conditions at the high latitudes of the Atlantic because of the increased pull of tropical Gulf Stream waters, and vice versa. Scientists have measured past variations in AMOC strength and correlated them to Northern Hemisphere terrestrial climate, such as during the short-term reversals to glacial conditions of the Younger Dryas around 12,000 years ago (https://www.pnas.org/doi/10.1073/pnas.1207381109) and the 8200-year cold event (https://tos.org/oceanography/article/rapid-climate-change-and-climate-surprises-a-look-back-and-ahead). In both of these cases, an influx of fresh water from glacial ice melting into the North Atlantic had a freshening effect on the surface ocean, temporarily reducing its density, slowing down the export of deep waters, and jamming up the AMOC.

Fast forward several thousand years from these geologic events, and we seem to be witnessing an AMOC slowdown again, but this time from human impacts (https://www.severe-weather.eu/global-weather/gulf-stream-amoc-circulation-collapse-freshwater-imbalance-usa-europe-fa/#:~:text=But%20why%20is%20the%20Gulf,imbalance%20in%20the%20ocean%20current). And it seems to be the same old driver—increased meltwater from glaciers is again freshening the waters of the surface North Atlantic. The sources of this meltwater are multiple, but perhaps nothing captures the extent and rate of glacial melting more than Greenland. The vast, frozen expanse that used to be a temporary home to Vikings when the climate was a little warmer might soon be quite hospitable again, this time owing to climate change and the rapid melting of the Greenland Ice Sheet. Greenland ice holds 2,850,000 cubic kilometers of freshwater, but perhaps a better and more frightening visual is that if the entire ice sheet would melt, global sea levels would rise by about 7 m. That 24 feet of sea level rise would swallow many coastal cities and communities, and a few entire nations. Future sea level rise and coastal response will be discussed later, but what does the slowdown of the AMOC mean for the climate? The geologic record paints a very grim picture.

As the investing phrase famously states, "past performance is not a guarantee of future returns." In the case of climatic shifts, the past performance of a meltwater-slowed AMOC was a temporary but significant regional cooling on land, based on a host of ice core and terrestrial records. This cooling was largely confined to the North Atlantic region, hitting Scandinavia and Northern Europe swiftly and hard, sending glaciers raging back down alpine slopes and creating glacial-like conditions that persisted for centuries. There is also evidence, from the 8200-year event at least, of these glacial conditions sweeping all the way around the high latitudes of North America, based on lake sediment cores from the Coast Mountains of western Canada (https://tos.org/oceanography/article/rapid-climate-change-and-climate-surprises-a-look-back-and-ahead). So, ironically, during a past period of rapid global warming from Ice Age to Holocene conditions, some confined regions temporarily went the opposite direction. A key question, based on these past events, is will the current climate-slowed AMOC wreak similar havoc in parts of the Northern Hemisphere?

Any fan of apocalyptic CliFi will certainly recognize this story line, as it was central to the scenario imagined in the Hollywood movie "The Day After Tomorrow." Fig. 9.6 Here, a handsome paleoclimatologist (aren't we all?) brandishes a climate record of the 8200-year event in front of the US Congress in a call for climate action, as evidence was mounting that the AMOC was slowing again. Up to that point in the film, it was a masterpiece of scientific accuracy, but it went downhill soon after. Granted, it is hard to keep a theater audience rapt for the 150 years or so that such a slowdown would play out, and the visuals of slightly colder conditions causing some minor disruptions in food production wouldn't stack up to the film's images of ice sheets roaring down the streets of New York City like freight trains.

Figure 9.6 ***Advertisement for the CliFi Film "The Day After Tomorrow"*** This climate apocalypse film envisions the chilling impacts of a slowdown of the Atlantic Meridional Overturning Circulation pattern from climate change on the Northern Hemisphere.

Nevertheless, the film portrays an extreme end member of the concern that scientists have related to an AMOC slowdown. Meltwater from Greenland and other proximal glaciers is causing the one area of cooling blue in an otherwise deep red map of global ocean warming, but this hasn't translated into cooler conditions over Europe—in fact, the opposite is the case, likely because the forces of warming have far exceeded the cooling that previously occurred as the Earth was transitioning from an Ice Age to a Holocene climate. The one difference is that the baseline has changed, and in this case, the warming and AMOC slowdown is occurring on an already warm planet, and thus the AMOC might have less capacity to substantially alter regional climate than it had before.

Threats at the poles

Polar regions are warming faster than any other place on the planet. This is partly due to enhanced heat transfer from the equatorial regions to the poles, along with the fact that the tropics are nearer to their thermal limits than the poles, where the positive feedback between ice loss, lowered albedo, and increased local temperatures further amplify polar warming. Polar warming is causing a significant loss of the planet's cryosphere and the range of biogeochemical and ecological systems that exist there. Many of the biogeochemical systems in the warming polar regions are destined to drive a host of positive feedbacks that will make climate change worse, or even a lot worse. One that has received significant attention is the so-called "Climate Bomb," a version of which is called the "Clathrate Gun." Both of these militaristic terms refer to the very real fact that there are significant quantities of carbon locking into the soils and peats, and under the ice, of the polar regions (https://www.nature.com/articles/d41586-021-00659-y). This carbon is stored in the large peat deposits that naturally form in this environment, along with the frozen soils and various deposits of methane-rich clathrate. These materials are all very vulnerable to warming conditions, and ultimately a fully warm Arctic (and Antarctic) would produce a tremendous amount of carbon in the atmosphere, exacerbating the very process of climate change that set this situation up in the first place. Many scientists argue that the "Climate bomb" might be more like a "Climate slow fizzle," at least in the short term. Based on field evidence and climate models, carbon currently released from the warming Arctic seems to be somewhat offset by polar sinks, and the net balance is a little extra carbon being released, but not a lot. This might change when you look farther in the future, particularly if that future is significantly warmer than most policymakers hope that it will be.

Land-based glacial ice is melting at an alarming rate, but many polar habitats, and animals like polar bears, depend on sea ice cover. Arctic sea ice is in serious decline. In 2021, for example, the summer minimum Arctic sea ice cover extended over 4.72 million square kilometers (1.82 million square miles)—this constitutes a 25% decline in sea ice cover from its 1981—2010 average (https://www.climate.gov/news-features/understanding-climate/climate-change-arctic-sea-ice-summer-minimum) (Fig. 9.7). Furthermore, the ice that is there is increasingly thin and not the thick, multi-year ice that used to dominate the Arctic. This sea ice thinning has basically opened the Arctic up for business, as now it will soon for viable to transport goods from some countries through the fable Northwest Passage at a fraction of the cost and time that a route through the Panama Canal would take. It also makes prospects for offshore oil production and seabed mining much more feasible. Of course, expansion of any oil exploration and exploitation just makes the climate change situation worse (its own strange positive feedback here), but all of these shipping and operations also make search-and-rescue efforts more challenging given the lack of shipping support infrastructure in this area. It also opens up another border to protect, which was previously protected through this nearly impenetrable blanket of ice.

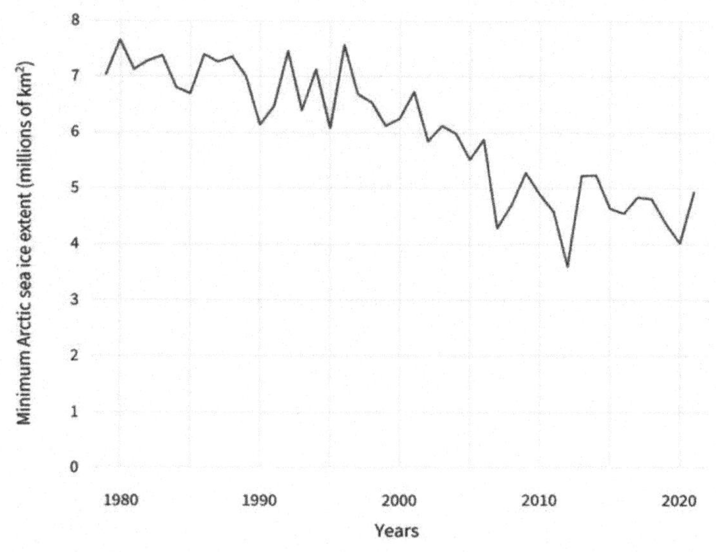

Figure 9.7 Arctic Sea Ice extent The extent of summer Arctic sea ice has declined sharply since 2000.

Although for now, at least, the cataclysmic release of polar carbon might not be the most pressing concern that we face, the loss of cryospheric habitat very well might be for plants and animals. The starving polar bear stranded on a bit of drifting sea ice is an iconic image of climate change (Fig. 9.8). And there is no question that the hunting habits of polar bears, and the ranges of many polar organisms, are changing rapidly. Polar bear's primary habitat is sea ice, from which they hunt ringed seals, and in some areas of its range sea ice cover is declining. But even here, ecosystems seem to display some level of resilience, and various surveys of polar bear populations indicate that they are actually quite robust (https://www.carbonbrief.org/polar-bears-and-climate-change-what-does-the-science-say/). They are adopting different patterns of travel and hunting, to be sure, as their typical hunting grounds change with the substantial changes in sea ice cover. There is no question that some polar bear groups are in serious decline, but not all of them, and there is hope that populations can shift or adapt to accommodate habitat changes at least to a certain extent. If the future holds consistently ice-free summers, this resilience window might be exceeded.

Figure 9.8 The starving polar bear stranded on a bit of drifting sea ice is an iconic image of climate change.

Threats on land

Let's talk about the birds and the bees. No, this is not an awkward discussion of the reproductive process but instead one of significant biotic loss. The populations of birds, bees, and flying insects, in general, have sharply declined in some parts of the world, which is a troubling harbinger of the full impacts that humans have had on the landscape. These biotic declines are not just from climate change but stem from multiple compounding pressures such as changes in landscape forage, the night lighting of cities and attraction of migrating animals to these nutrition-poor areas, application of myriad chemicals to control pests and weeds in cultivated areas, and transmission of diseases compounded by food source locations (like trash cans and bird feeders). The numbers are bad, including a survey that revealed that 3 billion birds are no longer present in the USA and Canada compared to 1970 (https://www.science.org/doi/10.1126/science.aaw1313#:~:text=Integration%20of%20range%2Dwide%20population,a%20recent%2010%2Dyear%20period, https://www.birds.cornell.edu/home/bring-birds-back/#:~:text=An%20alarming%20new%20study%20reveals,span%20of%20a%20human%20lifetime). Yes, you read the billion correctly, and those losses range across biomes, whether forest or grassland. A shocking 90% of the losses have been in only 12 families of birds, many of them the common birds, including dark-eyed juncos, white-throated sparrows, meadowlarks, and blackbirds. Pollinators like bees and other flying insects are also in substantial decline. The bee loss linked to Colony Collapse Disorder (https://www.epa.gov/pollinator-protection/colony-collapse-disorder), raises some concern for entomologists and farmers alike, as bees are critical for pollinating many fruit and nut tree crops. Flying insects meanwhile have suffered a decrease in biomass of 75% since 1990 (https://journals.plos.org/plosone/article?id=10.1371/journal.pone.0185809), the result of multiple chemical, habitat, and climate insults. The loss of birds and flying insects are not just documented in detailed scientific studies but are clear on a personal level as well—the days of an insect-crusted car windshield after a road trip are a thing of the past, and seeing vast clouds of birds over fields and forests is more and more rare.

The signature of climate change on land is easily discernible, through thermometers and rain gauges alike. Most areas of the planet are getting warmer, and many are getting MUCH warmer. Precipitation patterns are also changing, with an observed tendency for dry places to get drier, and wet to get wetter. Extreme end-members of precipitation are also extended, with multi-year drought becoming a common feature of many areas, and extreme precipitation and flooding in others. These extended extreme weather events are consistent with our ever-evolving understanding of climate change impacts on the atmosphere, which can drive more stable and long-lived high and pressure zones. These stalled high- and low-pressure regions bring extended heat waves and drought (for high-pressure zones) and storms and extreme precipitation (for low-pressure zones). Even freak ice storms in far southern latitudes, like that which struck deep into Texas in 2021 and crippled its energy grid (https://www.texastribune.org/series/winter-storm-power-outage/), are consistent with climate change, as they are related to a weakening polar front from higher polar ground temperatures and occasional intrusions of Arctic air into the temperate zones. Such is weather on a warming planet.

Land ecosystems are strongly dependent on climatic stability, and the relative instability due to climate change is impacting the distribution of forests, grasslands, and species. But it is difficult to understand the full extent of climate change impacts in these systems given the myriad other ecosystem alterations that humans are responsible for. Nearly all of our planetary ecosystems have "felt" human influences beyond climate change, and most of them are completely altered through deforestation, farming, the introduction of nonnative species, and development. Humans have reduced, dissected, and fragmented many ecosystems beyond the point of viability in some cases. But beyond these other influences, climate change has had several direct and indirect impacts on terrestrial ecosystems that are not often discussed, and have alarming potential impacts on humans.

Carbon dioxide, junk food, and poison ivy

Increased concentrations of carbon dioxide in the atmosphere have a number of direct effects on plants, and the animals that interact with them. First, high carbon dioxide is turning all of the animal forage into junk food. OK, a bit of hyperbole there, but the basics are true. The ratio of carbon dioxide to various protein components of plants is not absolutely fixed. Think of plant carbon, derived from carbon dioxide during the photosynthetic process, to be sugar, which is exactly why plants create it. The nitrogen and the various amino acids and proteins are other products of photosynthesis, and are used for basic biological activities like stem and leaf growth. All things being equal, when plants are grown in higher carbon dioxide conditions, the ratio of carbon to those other components go up. So the product of photosynthesis becomes effectively a greater sugar: protein ratio in the plants. Although this doesn't affect the plants themselves in a significant way, it does affect the animals that eat that plant matter. Plants become more sugar-filled and less protein-rich per gram—i.e., more junky (https://globalhealth.washington.edu/news/2019/04/23/high-co2-levels-will-wreck-plants-nutritional-value-so-don-t-plan-surviving). And if your parents taught you one thing, it is that junk food is bad for you! This shift in plant nutrition is a major concern for wildlife biologists who see this impacting the general health of foraging animals.

Another direct impact of carbon dioxide on plants is that at higher concentrations, it generally increases the productivity of plant species, particularly the C3 plants (https://www.nature.com/scitable/knowledge/library/effects-of-rising-atmospheric-concentrations-of-carbon-13254108/). The C4 photosynthetic pathway of many types of grass and crops like corn evolved to concentrate carbon dioxide in response to the decrease in atmospheric carbon dioxide in the latter part of the Cenozoic, and thus they seem to see no benefit from higher carbon dioxide concentrations. These measurements of increased plant growth have been conducted in greenhouses and controlled field experiments where carbon dioxide is intentionally injected into the air around plants, including trees (which use the C3 pathway), and are supported by a general increase in "greening" of the terrestrial biosphere using satellite measurements. As noted above, the increased growth comes at the cost of decreased plant nutrition and probably does very little to increase the terrestrial sink for carbon dioxide considering all of the other disturbances that we have created to carbon sinks and the extremely high rate of current carbon emissions.

Several plant species do particularly well in higher carbon dioxide conditions, including a host of weed species that we consider nuisances. Poison ivy, dandelions, and ragweed absolutely love higher carbon dioxide conditions. We, humans, tend to absolutely loathe those aforementioned species. Well, we should leave dandelions out of the loathing category as they are beloved by children who help disperse them widely through wish-making and they seem only to be a nuisance for those loving an orderly monoculture lawn. Increased carbon dioxide has boosted the growth of poison ivy Fig. 9.9, and substantially increased the amount of urushiol, the toxic component that causes skin misery (https://www.webmd.com/skin-problems-and-treatments/features/climate-change-brings-super-poison-ivy#:~:text=As%20the%20level%20of%20carbon,and%20miserable%20poison%20ivy%20rash). And that poison ivy toxin is not just more plentiful, is also more powerful. Leaving the ambivalent dandelion aside, ragweed also flourishes in higher carbon dioxide environments. Although likely benign to most organisms, the pollen from ragweed wreaks havoc in human sinuses and causes severe seasonal allergies that can be deadly. And in a higher carbon dioxide world, ragweed pollen is especially virulent and causes human lung inflammation (https://onlinelibrary.wiley.com/doi/10.1111/all.14618). In these ways, the major climate change driver carbon

Figure 9.9 ***Poison Ivy*** Poison ivy is one of several "weed species" that grows better in the newer carbon dioxide-rich atmosphere.

dioxide also plays a role in various other biological processes, some of which are harmful to humans.

Ticks, mosquitoes, and COVID-19?

Warmer winters are a blessing to many, even if wetter flooded springs are not. But the combination of warm winters and wet springs is driving a perilous shift in ecosystems in some parts of the world, with implications for human health. In temperate climates, cold winters tend to cut down on pest populations. Insects like mosquitoes and ticks are partially killed off during cold winters, reducing their virulence in warmer summer months. This climatic control is important for human health at least, as both of these insects harbor potential viral diseases that can be dangerous to people. Mosquitoes are a vector for any number of diseases, including avian flu, Zika, Dengue, Malaria, and Chikungunya. Ticks, on the other hand, can harbor and spread diseases of the rickettsia type—i.e., Lyme Disease. Warmer winters mean less die-off of these pests, and wetter springs mean more growth habitats for them.

Mosquitos have been called the "most dangerous animal in the world" (https://www.cdc.gov/globalhealth/stories/2019/world-deadliest-animal.html) given how many people have died from these diseases throughout history. It is worth noting that mosquitoes are not just a human menace—they have been around since the time of the dinosaurs, and likely caused them grief as well (https://www.amazon.com/Mosquito-Human-History-Deadliest-Predator/dp/0735235791). Warmer winters mean less die-off of these pests, and wetter springs mean more growth habitats for them. In some cities of the USA, for example, mosquitoes numbers have increased by four-fold Fig. 9.10, and some of the particular mosquito varieties that can carry zika and dengue, the Aedis Aegypti, are now endemic (https://link.springer.com/article/10.1007/s10584-020-02710-9). For now, at least, these deadly parasitic pests seem to be gaining an upper hand. But help might be at hand, at least for mosquitoes. A vast array of research programs have shown that potentially deadly mosquito populations can be controlled by the introduction of genetically modified versions of the same mosquito, capable of breeding but being completely sterile and thus incapable of reproduction. But mosquito populations are actually relatively easy to control—by limiting the areas where ponded water can persist, allowing mosquito larvae to mature, urban areas can keep a lid on mosquito infestations.

Covid-19 is complicated, not by the science of this virus, but by the social and political conversations that have sparked around it. Conspiracies have been rampant—a hoax! Invented by the Chinese to cause the downfall of the western world! An accidental release from a controlled facility in Wuhan, China! All of these have proven false. COVID-19 was sparked as earlier coronaviruses were—by a zoonotic transfer of viruses. In this case, it was a transfer that caused a once-in-a-century disruption of a global society. But its spark in the forests of Asia was in fact coupled with

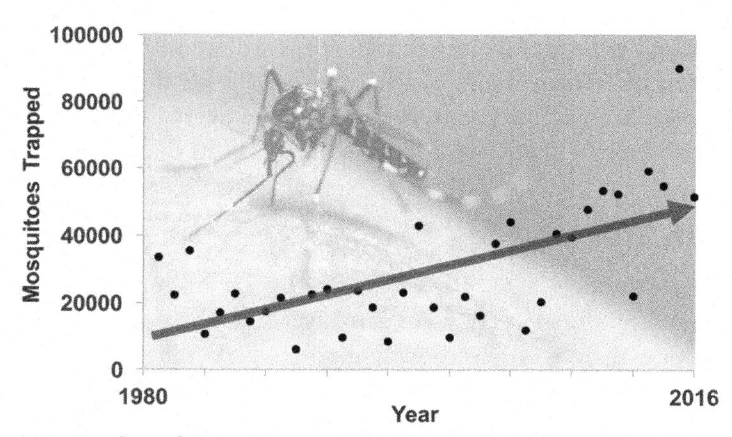

Figure 9.10 ***Number of Ticks Captured in Indianapolis, Indiana (USA) from 1980 to 2016*** The population of mosquitoes has increased substantially in many cities around the world, which poses threat to public health.

environmental, and climatic change. Scientists have long warned that ecological disturbance, the increased exploitation and loss of forest ecosystems, and the carving out of the forest for livestock production pose a significant risk to public health (https://www.pnas.org/doi/10.1073/pnas.2023540118). The explosion of COVID-19 around the globe proved them right. Zoonotic diseases are those passed between an animal host and a human. Like SARS-1 before it, SARS-2 (COVID-19) is a coronavirus vaguely related to the common cold, but clearly much more transmissible, and deadly.

These types of zoonotic disease outbreaks have become more frequent, and the elevation between an outbreak, an epidemic, and a pandemic is more likely given global travel. In the case of COVID-19, the disease was sparked by climatic and environmental changes that destroyed habitat and brought wildlife much closer to human contact. The disease vector was likely a bat, which carried the virus from its original wild host to domesticated animals, which were then brought diseased to a market in Wuhan, China, with a high number of people. The close contact between humans and diseased animals allowed this particularly transmissible virus to make the leap to us, and the rest, as they say, is history. Experts predict that these events will continue to pose significant global health risks (https://www.npr.org/sections/coronavirus-live-updates/2020/07/06/888077232/u-n-predicts-rise-in-diseases-that-jump-from-animals-to-humans), warranting changes in land use practices to avoid exacerbating wildlife incursion and of course, better surveillance and treatments. The scientific miracle that was enabled by global cooperation and delivered a ridiculously effective vaccine less than a year after identifying the virus will be touted as one of the greatest scientific achievements of the century. But we don't want to have to rely on this kind of cooperation, scientific discovery, and good fortune to confront all of the predicted zoonotic outbreaks in our future.

Earth 2100

Climate models have proven uncannily accurate at predicting the physics of future climate. They are actually MUCH easier to run and far more accurate than weather models, upon which the basics of climate models are built. Weather models take into account well-known factors that affect heat transport and moisture characteristics in a three-dimensional atmosphere, and their output is phenomenally accurate if one were planning to have a picnic today. But so many micro variables exist in atmospheric dynamics that have substantial and unpredictable ripple effects that these models are not always perfect at forecasting whether you should have a picnic next week. Climate models, on the other hand, have a simpler atmospheric structure and are sensitive to different types of variables than weather models. Much like it is easier to predict that you are more likely to pull off a successful picnic in the middle of summer than the middle of winter, climate models similarly reproduce average temperature and precipitation remarkably well. There are more sophisticated applications of climate models, including those that "down-scale" predictions to the state or even city level, which are constantly being developed. This will prove extremely valuable for a city planner in, say, Des Moines Iowa to gauge what scale of stormwater infrastructure needs to be built to control flooding in 2050 (more about this later) or in Miami to estimate storm surge levels for levee design. The standard climate models not only tend to agree with each other, but they can also be spun forward from a starting point of the 1990 climate and predict quite well what the 2022 climate is looking like.

There is one huge variable that confounds the reliability of all of the models in accurately predicting future climate. It is not the physics of the models themselves, which is quite robust, but rather us, and our collective choices. We are at an interesting inflection point, armed with substantial information, but facing a daunting challenge. We can reliably measure how much greenhouse gasses are emitted into the atmosphere per year, down even to the city scale. We also can reliably predict the future trajectory of climate change given those annual emissions. These are the easy part. The challenges are multiple, and not easy to reconcile. First, we cannot all agree on the need for immediacy in reducing and/or eliminating greenhouse gas emissions. Is it really the issue of our lifetimes, or are other pressing issues, like conflict, hunger, and the economy, more primary focuses at the moment? Second, we cannot all agree on what a transition would look like, who pays for it, and how it is disseminated around the globe. Various climate talks and agreements have occurred since climate change was raised as a potential future issue in the 1980's. Meanwhile, greenhouse gas emissions have continued ticking up as if none of those agreements existed, and even now, when the climate is not a future issue but a now issue, massive sectors of our economy still rely on burning fossil fuels. Indeed, the embedded infrastructure for fossil fuel surrounds us, from the countless gas stations to the pipelines delivering natural gas to our homes to the delivery trucks and trains and ships transporting our goods to the vehicles and airplanes that propel us from here to there. Up to this point, the transition to nonfossil fuel lifestyles has been incremental in the developed world, and virtually nonexistent in the developing world (although this latter group shares very little of the global carbon burden that has been emitted to the atmosphere since the Industrial Revolution).

A rapid global awakening to the climate crisis is not likely to change the carbon course immediately, but there have already been significant increases in the proportion of energy that is now produced by nonfossil fuel processes, like solar photovoltaic and wind generation. And more renewable production is being added every year. So, what does this mean for our projections of future carbon emissions, climate change, and ecosystem responses? Many scientists and economists see a mixed future, with the worst-case climate scenarios that many had dreaded a decade ago not likely to be in our near future, and the very best ones no longer achievable as we have already blown through a lot of the allowable carbon budget to achieve these. The window of realistic climate futures has narrowed, which means that we can similarly project the key climate factors that will likely affect global ecosystems, like temperature, precipitation, and sea level rise, with some confidence.

The more optimistic side of the climate window calls for rapid decarbonization of the electrical, transport, and manufacturing sectors and substantial growth of negative emissions to be achieved by the year 2050 (technically, this scenario is captured in the Representative Concentration Pathway (RCP) 2.6) Negative emissions are basically ways to suck carbon back out of the atmosphere, either through the good old fashion way of biological sinks like trees or technological ones like Direct Air Capture. Forests will be easier than technologies because we know how to grow trees at scale, but not necessarily how to suck carbon out of the atmosphere at scale. Having said that, though, there is a functional limit to how much forest growth can be built into the negative emissions equation because not all landscapes and climates are amenable to sustained forest growth. By whatever means manageable, this optimistic side of the climate window will yield climatic conditions in the year 2100 that will be a bit less than 2°C warmer than Pre-Industrial conditions Fig. 9.11. Pre-Industrial temperatures are typically used as the climate baseline, and for reference, as of 2022 global average temperatures are 1°C warmer than this, and thus the optimistic projections have the planet warming less than 1° more by the end of the century. The projected total sea level rise in this optimistic scenario is about 40 cm.

With global conditions, 1° warmer than today, and sea levels 40 cm higher, even the optimistic climate projection yields substantial impacts on life on the planet. What is somewhat hidden in these global averages is that ecosystems, and indeed people, don't survive in a theoretical global

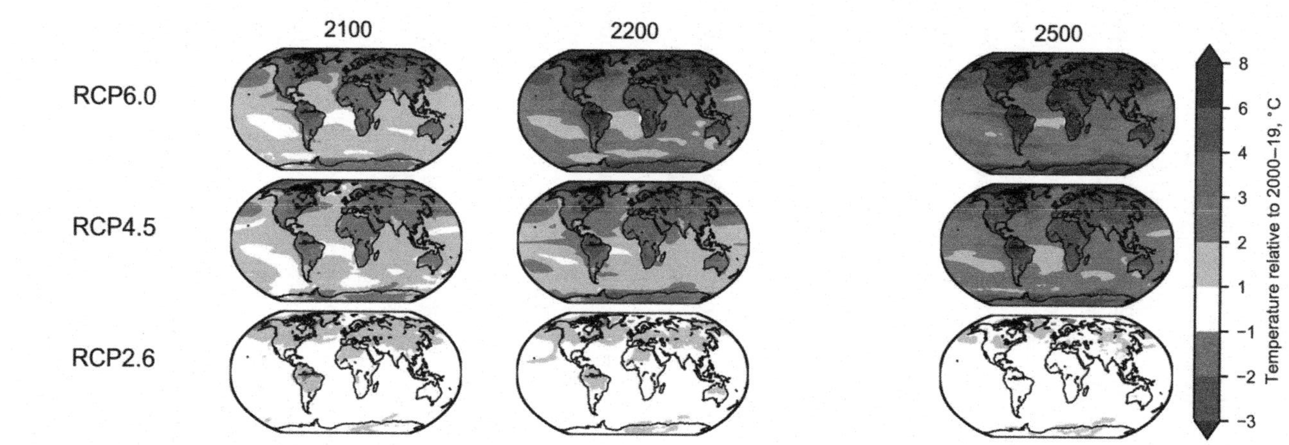

Figure 9.11 ***Climate Projections for 2100, 2300 and 2500*** Global temperature patterns are projected for the climate in 2100, 2300, and 2500 for lower (top) and higher (bottom) scenarios for carbon dioxide concentrations.

average state but rather in the climatic conditions where they are. A regional projection of this climate future shows a substantial increase in polar temperatures, and significant coastal impacts on low-lying areas and the ecosystems that exist there. Even this low-end projection of 2100 climate is not an optimistic one for some ecosystems like coral reefs that are already at the knife's edge of survival. To further examine the optimistic climate scenario, it is also useful to project beyond the year 2100. By 2300, temperatures and sea level are projected to be about what they were in 2100, as they will with an even further timeline of 2500. These projections show that if humanity exerts the effort to quickly turn the carbon ship around, in the same kind of global attention that was paid to eliminating sources of CFCs that were destroying the stratospheric ozone layer and finding a vaccine for COVID-19, we would have truly dodged the climate bullet. Global ecosystems would still be forever altered, and it is reasonable to assume that this interval of human dominance would still qualify as the sixth mass extinction. But our great-great-great-great grandchildren would live on a planet with a more stable and manageable climate.

In this "good news first" storyline, the climate crisis can be averted, but it will take substantial, systemic changes and planning to achieve this climate solution. Furthermore, the climate will still continue to change for a bit, and our own infrastructure planning will have to include this change. For example, 40 cm of sea level rise is not at all a trivial amount for low-lying coastal plains and the cities that have grown up on them, some of which are technically below sea level already. Dewatering and coastal armoring are one type of infrastructure response in this scenario, but coastal retreat is also a reasonable response, because storm surges and more powerful hurricanes may tax even the best-designed coastal armoring systems, and future infrastructure investments should probably be located where the future coastline will be, rather than solely protecting against a rising tide. Additionally, high temperatures and stormwater issues will plague many cities and towns in even the optimistic scenario. For example, Indianapolis, Indiana (USA) is undergoing a massive stormwater infrastructure upgrade, as are most other cities in the USA, to improve water quality and public health. The upgrade was designed to a fixed capacity to a climatic baseline set in the year 2000. Since then, extreme precipitation has increased by 15% and is expected to increase by another 15% by the year 2050 in the optimistic climate projection. Climate change was clearly not built into the design equations, and thus optimal infrastructure function will not be achieved. This is but one of many examples of fixed infrastructure not truly adding resilience against future change.

Now for the "bad news" climate scenario. As noted earlier, our current greenhouse gas emission trajectory is better than previously feared, and thus the more pessimistic future is a less bad one than it could be. But it still yields a pretty drastically altered global climate, as captured in RCP4.5. In this scenario, significant decarbonization of our energy systems is slow to roll out and negative emissions are not prioritized. This scenario yields a 2100 average global temperature increase of about 2°C above Pre-Industrial and a sea level rise of about 47 cm. Although the extra sea level rise from this scenario is only about 7 cm higher than from the optimistic scenario, the temperature increase is nearly double. By the year 2300, this becomes a temperature of about 3° warmer than Pre-Industrial and a sea level rise of over 80 cm. The more pessimistic climate scenario, featuring delays in getting carbon emissions under control, paints a very grim picture for global ecosystems, and humans, in this very long view. Of course, 2500 is very far away, and it is not wise to apply the same optics of technology and perceived limitations that far into the future. After all, who in the 1600's would have projected air conditioning, air travel, cell phones, or microwave popcorn?

Summary

The planet's evolutionary arc is one shaped by climate, and sometimes molding climate itself. Humans are the current heir to a lineage that started with algae, but we do not live in isolation from our fellow planetary travelers. Our bodies, and our societies, are dependent on the complex ecological web around us. As human activities and related climate change have chiseled into that web, species by species, our safety net has become more and more tattered. Many of the novel medicines and antibiotics that now save millions of lives were not cooked up in a beaker on a laboratory bench, but rather were extracted from natural species and organisms and then refined for our benefit. As we lose more and more species, do we even appreciate what we are losing? It could be a cure for cancer, depression, or genetic disease. But it could just as easily be an organism in the web that plays a vital role for another organism, which we do not even know about. The disturbance to the ecological web helped to propagate a global pandemic, and there is no reason to think that a repeat is not in our future if we continue eroding our ecological safety net.

A wise course is to tread lightly, including ensuring that the least painful climate future is ahead of us. We are, after all, in the climate driver's seat. The physics are easy—more greenhouse gas emissions and higher atmospheric concentrations mean a warmer planet. This situation has played out throughout Earth's history, and the rules haven't changed just because it is us producing the greenhouse gasses. The two likely climate scenarios laid out as windows into the different futures are only reasonable guesses based on the current situation. There are better, and worse, ways that it might all turn out. But even those two likely scenarios paint a stark contrast between a more pleasant, equitable, and ecologically robust climate future, and one that is not so nice. It seems that the choice is clear, but will we make the right one? This seasoned planet that has sustained life for almost all of its billions of years of existence will eventually shrug us off as a brief but disruptive hiccup, but for now, while we inhabit this beautiful orb and have some measure of free will, let's make the right choice.

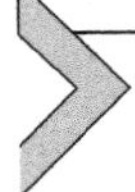

Did you know that?

Human psychology is strongly behind why some downplay or even doubt, the scientific realities of climate change

As a climate scientist myself, I was raised by the assumption that more facts on a topic yield a more convincing argument, which then would change minds. For example, by showing the trajectory of greenhouse gas concentrations and ice loss to a general audience, I would assume that audience would then be enlightened. If any doubts remained, more facts would certainly sway their opinions, right? No, absolutely wrong. In fact, studies on climate change perceptions have shown that providing more data to people who are not convinced that human-produced climate change is occurring actually makes them even less convinced about climate. And distressingly, doing the same to an audience who is not well versed in the science but who "believe" in climate change has the same effect–they come out of the conversation less convinced that climate change is occurring. What in the world is happening here?

Psychologists and sociologists have explained that we as human animals are not altogether rational organisms (big surprise, right?). For many reasons, we use our own life experiences to frame our understanding of the world around us. Yes, we feel bad about the poor starving polar bear stranded on the ice sheet, but the climate story being told there is just theoretical–until someone gets washed out of their home in a climate-fueled flood, it remains a theoretical construct that requires no action. The many stories of people who refused COVID-19 vaccines but "see the light" when they are laid out on a hospital bed with a bad case of COVID and urge their friends and family to get vaccinated are evidence of this distance effect. We also suffer from confirmation bias, basically only believing the tiny morsels of facts that fit our worldview and discounting the mountains of facts that do not. This also has played out in recent years, with the false claim that the 2020 US Presidential elections were somehow rigged, thoroughly discounted by a mountain of facts, still swaying a large fraction of the population who latch onto cherry-picked observations and stories even as opposing facts have piled up.

These sociological factors have inspired an entirely new set of climate communicators, including Katharine Hayhoe and Marshall Shepherd, and Michael Mann. These communicators understand that connections have to be made first with the audience, some common ground reached, before a conversation can be fruitful. And they also understand the power of the story and tap into that lived experience to tell that story. Take for example discussing climate change with a rural farmer in Iowa. For various ideological and political reasons, that farmer is likely to react negatively if the first words out of a climate scientist's mouth are "climate change." Instead, it is most fruitful to start with a conversation about the weather. Everybody experiences weather–it is concrete, and of course, everyone likes to talk about it. If you start a conversation with the Iowan farmer about what winters were like when he or she (mostly he in the case of Iowa farmers) were growing up, they would describe bitter, snow-piled landscapes where every winter they would have to shovel 10 feet of snow out of the doorway just to get outside. Then you ask them what winter is like now, and they will note that it hardly snows at all anymore, and when it does, it doesn't stick around. In their own way, they are telling the story of climate change without being beaten over the head with plots and facts, and they are more likely to engage in the next part of the conversation about the need to be prepared for continued "weather" changes.

There are still some valid uncertainties in how well climate models accurately capture the potential range of impacts on the ground

Several uncertainties in the application of climate models for assessing future impacts are front and center in the minds of climate scientists and are areas of active research. One of these revolves around climate sensitivity to atmospheric carbon dioxide concentrations (https://www.carbonbrief.org/explainer-how-scientists-estimate-climate-sensitivity/). In broad strokes, we understand what a certain increase of carbon dioxide in the atmosphere will yield in terms of temperature change, based largely on how these factors have interacted in the past. Because we can directly measure carbon dioxide concentration in the past atmosphere from the air bubbles trapped in ice, and independently determine the temperature of the air above that ice, we have a reasonable understanding of the carbon dioxide/temperature relationship.

But there are fewer of these types of accurate direct correlations outside of ice environments, which are mostly restricted to polar and high-altitude settings. Thus, our global picture does not include such accurate information from tropical or equatorial regions. There is only a relatively narrow range in these sensitivity estimates compared to the substantial projected increases in atmospheric carbon dioxide, but this one basic aspect of climate model sensitivity certainly keep modelers up at night!

Another uncertainty that has potentially profound impacts on the future is the sensitivity and response time of ice melt to temperature increases. As discussed earlier, ice has self-protecting albedo properties, such that it is hard to get ice to start melting, but once it starts, it can melt very quickly. Fig. 9.12 This nonsymmetric response to temperature increases makes it difficult to know the potential range and timing of ice-melt-induced sea

Figure 9.12 ***Larsen Ice Shelf breaking Apart in Antarctica*** Large ice shelves had collapsed rapidly in glacial areas, and the resulting melt and sea level rise are not currently well-accounted for in climate models.

level rise. This challenge is compounded by the fact that the stability of many of the major glaciers from which melting would occur is not particularly well-understood. If, for example, a large section of the marine portion, or tongue, of a glacier breaks off from its land-based upstream section, that chunk of ice will melt rapidly into the surface ocean.

This type of ice sheet collapse occurs due to a number of factors that are difficult to measure and thus predict, including where the underside of the glacial ends is experiencing incursion of warmer ocean water, which speeds melting. These "ice sheet collapse scenarios" are not typically represented in the projections of sea level that the climate models predict, raising the specter that instead of a 40 cm sea level rise by 2100, the future might bring a much greater rate of sea level rise. If this situation plays out, many coastal communities will be hard pressed to build adequate climate resilience infrastructure to protect themselves.

Index

Note: 'Page numbers followed by 'f' indicate figures.'

I

J

L

M